T0021781

The History of Computing: A Very Short Introduction

VERY SHORT INTRODUCTIONS are for anyone wanting a stimulating and accessible way into a new subject. They are written by experts, and have been translated into more than 45 different languages.

The series began in 1995, and now covers a wide variety of topics in every discipline. The VSI library currently contains over 700 volumes—a Very Short Introduction to everything from Psychology and Philosophy of Science to American History and Relativity—and continues to grow in every subject area.

Very Short Introductions available now:

ABOLITIONISM Richard S. Newman
THE ABRAHAMIC RELIGIONS Charles L. Cohen
ACCOUNTING Christopher Nobes
ADOLESCENCE Peter K. Smith
THEODOR W. ADORNO Andrew Bowie
ADVERTISING Winston Fletcher
AERIAL WARFARE Frank Ledwidge
AESTHETICS Bence Nanay
AFRICAN AMERICAN RELIGION Eddie S. Glaude Jr
AFRICAN HISTORY John Parker and Richard Rathbone
AFRICAN POLITICS Ian Taylor
AFRICAN RELIGIONS Jacob K. Olupona
AGEING Nancy A. Pachana
AGNOSTICISM Robin Le Poidevin
AGRICULTURE Paul Brassley and Richard Soffe
ALEXANDER THE GREAT Hugh Bowden
ALGEBRA Peter M. Higgins
AMERICAN BUSINESS HISTORY Walter A. Friedman
AMERICAN CULTURAL HISTORY Eric Avila
AMERICAN FOREIGN RELATIONS Andrew Preston
AMERICAN HISTORY Paul S. Boyer
AMERICAN IMMIGRATION David A. Gerber
AMERICAN INTELLECTUAL HISTORY Jennifer Ratner-Rosenhagen
THE AMERICAN JUDICIAL SYSTEM Charles L. Zelden
AMERICAN LEGAL HISTORY G. Edward White
AMERICAN MILITARY HISTORY Joseph T. Glatthaar
AMERICAN NAVAL HISTORY Craig L. Symonds
AMERICAN POETRY David Caplan
AMERICAN POLITICAL HISTORY Donald Critchlow
AMERICAN POLITICAL PARTIES AND ELECTIONS L. Sandy Maisel
AMERICAN POLITICS Richard M. Valelly
THE AMERICAN PRESIDENCY Charles O. Jones
THE AMERICAN REVOLUTION Robert J. Allison
AMERICAN SLAVERY Heather Andrea Williams
THE AMERICAN SOUTH Charles Reagan Wilson
THE AMERICAN WEST Stephen Aron
AMERICAN WOMEN'S HISTORY Susan Ware
AMPHIBIANS T. S. Kemp
ANAESTHESIA Aidan O'Donnell
ANALYTIC PHILOSOPHY Michael Beaney
ANARCHISM Alex Prichard

ANCIENT ASSYRIA Karen Radner
ANCIENT EGYPT Ian Shaw
ANCIENT EGYPTIAN ART AND ARCHITECTURE Christina Riggs
ANCIENT GREECE Paul Cartledge
THE ANCIENT NEAR EAST Amanda H. Podany
ANCIENT PHILOSOPHY Julia Annas
ANCIENT WARFARE Harry Sidebottom
ANGELS David Albert Jones
ANGLICANISM Mark Chapman
THE ANGLO-SAXON AGE John Blair
ANIMAL BEHAVIOUR Tristram D. Wyatt
THE ANIMAL KINGDOM Peter Holland
ANIMAL RIGHTS David DeGrazia
THE ANTARCTIC Klaus Dodds
ANTHROPOCENE Erle C. Ellis
ANTISEMITISM Steven Beller
ANXIETY Daniel Freeman and Jason Freeman
THE APOCRYPHAL GOSPELS Paul Foster
APPLIED MATHEMATICS Alain Goriely
THOMAS AQUINAS Fergus Kerr
ARBITRATION Thomas Schultz and Thomas Grant
ARCHAEOLOGY Paul Bahn
ARCHITECTURE Andrew Ballantyne
THE ARCTIC Klaus Dodds and Jamie Woodward
ARISTOCRACY William Doyle
ARISTOTLE Jonathan Barnes
ART HISTORY Dana Arnold
ART THEORY Cynthia Freeland
ARTIFICIAL INTELLIGENCE Margaret A. Boden
ASIAN AMERICAN HISTORY Madeline Y. Hsu
ASTROBIOLOGY David C. Catling
ASTROPHYSICS James Binney
ATHEISM Julian Baggini
THE ATMOSPHERE Paul I. Palmer
AUGUSTINE Henry Chadwick
JANE AUSTEN Tom Keymer
AUSTRALIA Kenneth Morgan
AUTISM Uta Frith
AUTOBIOGRAPHY Laura Marcus
THE AVANT GARDE David Cottington
THE AZTECS Davíd Carrasco
BABYLONIA Trevor Bryce
BACTERIA Sebastian G. B. Amyes
BANKING John Goddard and John O. S. Wilson
BARTHES Jonathan Culler
THE BEATS David Sterritt
BEAUTY Roger Scruton
LUDWIG VAN BEETHOVEN Mark Evan Bonds
BEHAVIOURAL ECONOMICS Michelle Baddeley
BESTSELLERS John Sutherland
THE BIBLE John Riches
BIBLICAL ARCHAEOLOGY Eric H. Cline
BIG DATA Dawn E. Holmes
BIOCHEMISTRY Mark Lorch
BIOGEOGRAPHY Mark V. Lomolino
BIOGRAPHY Hermione Lee
BIOMETRICS Michael Fairhurst
ELIZABETH BISHOP Jonathan F. S. Post
BLACK HOLES Katherine Blundell
BLASPHEMY Yvonne Sherwood
BLOOD Chris Cooper
THE BLUES Elijah Wald
THE BODY Chris Shilling
NIELS BOHR J. L. Heilbron
THE BOOK OF COMMON PRAYER Brian Cummings
THE BOOK OF MORMON Terryl Givens
BORDERS Alexander C. Diener and Joshua Hagen
THE BRAIN Michael O'Shea
BRANDING Robert Jones
THE BRICS Andrew F. Cooper
THE BRITISH CONSTITUTION Martin Loughlin
THE BRITISH EMPIRE Ashley Jackson
BRITISH POLITICS Tony Wright
BUDDHA Michael Carrithers
BUDDHISM Damien Keown
BUDDHIST ETHICS Damien Keown
BYZANTIUM Peter Sarris
CALVINISM Jon Balserak
ALBERT CAMUS Oliver Gloag

CANADA Donald Wright
CANCER Nicholas James
CAPITALISM James Fulcher
CATHOLICISM Gerald O'Collins
CAUSATION Stephen Mumford and Rani Lill Anjum
THE CELL Terence Allen and Graham Cowling
THE CELTS Barry Cunliffe
CHAOS Leonard Smith
GEOFFREY CHAUCER David Wallace
CHEMISTRY Peter Atkins
CHILD PSYCHOLOGY Usha Goswami
CHILDREN'S LITERATURE Kimberley Reynolds
CHINESE LITERATURE Sabina Knight
CHOICE THEORY Michael Allingham
CHRISTIAN ART Beth Williamson
CHRISTIAN ETHICS D. Stephen Long
CHRISTIANITY Linda Woodhead
CIRCADIAN RHYTHMS Russell Foster and Leon Kreitzman
CITIZENSHIP Richard Bellamy
CITY PLANNING Carl Abbott
CIVIL ENGINEERING David Muir Wood
CLASSICAL LITERATURE William Allan
CLASSICAL MYTHOLOGY Helen Morales
CLASSICS Mary Beard and John Henderson
CLAUSEWITZ Michael Howard
CLIMATE Mark Maslin
CLIMATE CHANGE Mark Maslin
CLINICAL PSYCHOLOGY Susan Llewelyn and Katie Aafjes-van Doorn
COGNITIVE BEHAVIOURAL THERAPY Freda McManus
COGNITIVE NEUROSCIENCE Richard Passingham
THE COLD WAR Robert J. McMahon
COLONIAL AMERICA Alan Taylor
COLONIAL LATIN AMERICAN LITERATURE Rolena Adorno
COMBINATORICS Robin Wilson
COMEDY Matthew Bevis
COMMUNISM Leslie Holmes
COMPARATIVE LITERATURE Ben Hutchinson
COMPETITION AND ANTITRUST LAW Ariel Ezrachi
COMPLEXITY John H. Holland
THE COMPUTER Darrel Ince
COMPUTER SCIENCE Subrata Dasgupta
CONCENTRATION CAMPS Dan Stone
CONFUCIANISM Daniel K. Gardner
THE CONQUISTADORS Matthew Restall and Felipe Fernández-Armesto
CONSCIENCE Paul Strohm
CONSCIOUSNESS Susan Blackmore
CONTEMPORARY ART Julian Stallabrass
CONTEMPORARY FICTION Robert Eaglestone
CONTINENTAL PHILOSOPHY Simon Critchley
COPERNICUS Owen Gingerich
CORAL REEFS Charles Sheppard
CORPORATE SOCIAL RESPONSIBILITY Jeremy Moon
CORRUPTION Leslie Holmes
COSMOLOGY Peter Coles
COUNTRY MUSIC Richard Carlin
CREATIVITY Vlad Glăveanu
CRIME FICTION Richard Bradford
CRIMINAL JUSTICE Julian V. Roberts
CRIMINOLOGY Tim Newburn
CRITICAL THEORY Stephen Eric Bronner
THE CRUSADES Christopher Tyerman
CRYPTOGRAPHY Fred Piper and Sean Murphy
CRYSTALLOGRAPHY A. M. Glazer
THE CULTURAL REVOLUTION Richard Curt Kraus
DADA AND SURREALISM David Hopkins
DANTE Peter Hainsworth and David Robey
DARWIN Jonathan Howard
THE DEAD SEA SCROLLS Timothy H. Lim
DECADENCE David Weir
DECOLONIZATION Dane Kennedy
DEMENTIA Kathleen Taylor
DEMOCRACY Bernard Crick

DEMOGRAPHY Sarah Harper
DEPRESSION Jan Scott and Mary Jane Tacchi
DERRIDA Simon Glendinning
DESCARTES Tom Sorell
DESERTS Nick Middleton
DESIGN John Heskett
DEVELOPMENT Ian Goldin
DEVELOPMENTAL BIOLOGY Lewis Wolpert
THE DEVIL Darren Oldridge
DIASPORA Kevin Kenny
CHARLES DICKENS Jenny Hartley
DICTIONARIES Lynda Mugglestone
DINOSAURS David Norman
DIPLOMATIC HISTORY Joseph M. Siracusa
DOCUMENTARY FILM Patricia Aufderheide
DREAMING J. Allan Hobson
DRUGS Les Iversen
DRUIDS Barry Cunliffe
DYNASTY Jeroen Duindam
DYSLEXIA Margaret J. Snowling
EARLY MUSIC Thomas Forrest Kelly
THE EARTH Martin Redfern
EARTH SYSTEM SCIENCE Tim Lenton
ECOLOGY Jaboury Ghazoul
ECONOMICS Partha Dasgupta
EDUCATION Gary Thomas
EGYPTIAN MYTH Geraldine Pinch
EIGHTEENTH-CENTURY BRITAIN Paul Langford
THE ELEMENTS Philip Ball
EMOTION Dylan Evans
EMPIRE Stephen Howe
EMPLOYMENT LAW David Cabrelli
ENERGY SYSTEMS Nick Jenkins
ENGELS Terrell Carver
ENGINEERING David Blockley
THE ENGLISH LANGUAGE Simon Horobin
ENGLISH LITERATURE Jonathan Bate
THE ENLIGHTENMENT John Robertson
ENTREPRENEURSHIP Paul Westhead and Mike Wright
ENVIRONMENTAL ECONOMICS Stephen Smith
ENVIRONMENTAL ETHICS Robin Attfield
ENVIRONMENTAL LAW Elizabeth Fisher
ENVIRONMENTAL POLITICS Andrew Dobson
ENZYMES Paul Engel
EPICUREANISM Catherine Wilson
EPIDEMIOLOGY Rodolfo Saracci
ETHICS Simon Blackburn
ETHNOMUSICOLOGY Timothy Rice
THE ETRUSCANS Christopher Smith
EUGENICS Philippa Levine
THE EUROPEAN UNION Simon Usherwood and John Pinder
EUROPEAN UNION LAW Anthony Arnull
EVANGELICALISM John Stackhouse
EVOLUTION Brian and Deborah Charlesworth
EXISTENTIALISM Thomas Flynn
EXPLORATION Stewart A. Weaver
EXTINCTION Paul B. Wignall
THE EYE Michael Land
FAIRY TALE Marina Warner
FAMILY LAW Jonathan Herring
MICHAEL FARADAY Frank A. J. L. James
FASCISM Kevin Passmore
FASHION Rebecca Arnold
FEDERALISM Mark J. Rozell and Clyde Wilcox
FEMINISM Margaret Walters
FILM Michael Wood
FILM MUSIC Kathryn Kalinak
FILM NOIR James Naremore
FIRE Andrew C. Scott
THE FIRST WORLD WAR Michael Howard
FLUID MECHANICS Eric Lauga
FOLK MUSIC Mark Slobin
FOOD John Krebs
FORENSIC PSYCHOLOGY David Canter
FORENSIC SCIENCE Jim Fraser
FORESTS Jaboury Ghazoul
FOSSILS Keith Thomson
FOUCAULT Gary Gutting
THE FOUNDING FATHERS R. B. Bernstein

FRACTALS Kenneth Falconer
FREE SPEECH Nigel Warburton
FREE WILL Thomas Pink
FREEMASONRY Andreas Önnerfors
FRENCH LITERATURE John D. Lyons
FRENCH PHILOSOPHY
Stephen Gaukroger and Knox Peden
THE FRENCH REVOLUTION
William Doyle
FREUD Anthony Storr
FUNDAMENTALISM Malise Ruthven
FUNGI Nicholas P. Money
THE FUTURE Jennifer M. Gidley
GALAXIES John Gribbin
GALILEO Stillman Drake
GAME THEORY Ken Binmore
GANDHI Bhikhu Parekh
GARDEN HISTORY Gordon Campbell
GENES Jonathan Slack
GENIUS Andrew Robinson
GENOMICS John Archibald
GEOGRAPHY John Matthews and
David Herbert
GEOLOGY Jan Zalasiewicz
GEOMETRY Maciej Dunajski
GEOPHYSICS William Lowrie
GEOPOLITICS Klaus Dodds
GERMAN LITERATURE Nicholas Boyle
GERMAN PHILOSOPHY Andrew Bowie
THE GHETTO Bryan Cheyette
GLACIATION David J. A. Evans
GLOBAL CATASTROPHES Bill McGuire
GLOBAL ECONOMIC HISTORY
Robert C. Allen
GLOBAL ISLAM Nile Green
GLOBALIZATION Manfred B. Steger
GOD John Bowker
GOETHE Ritchie Robertson
THE GOTHIC Nick Groom
GOVERNANCE Mark Bevir
GRAVITY Timothy Clifton
THE GREAT DEPRESSION AND
THE NEW DEAL Eric Rauchway
HABEAS CORPUS Amanda Tyler
HABERMAS James Gordon Finlayson
THE HABSBURG EMPIRE
Martyn Rady
HAPPINESS Daniel M. Haybron
THE HARLEM RENAISSANCE
Cheryl A. Wall
THE HEBREW BIBLE AS
LITERATURE Tod Linafelt
HEGEL Peter Singer
HEIDEGGER Michael Inwood
THE HELLENISTIC AGE
Peter Thonemann
HEREDITY John Waller
HERMENEUTICS Jens Zimmermann
HERODOTUS Jennifer T. Roberts
HIEROGLYPHS Penelope Wilson
HINDUISM Kim Knott
HISTORY John H. Arnold
THE HISTORY OF ASTRONOMY
Michael Hoskin
THE HISTORY OF CHEMISTRY
William H. Brock
THE HISTORY OF CHILDHOOD
James Marten
THE HISTORY OF CINEMA
Geoffrey Nowell-Smith
THE HISTORY OF COMPUTING
Doron Swade
THE HISTORY OF LIFE
Michael Benton
THE HISTORY OF MATHEMATICS
Jacqueline Stedall
THE HISTORY OF MEDICINE
William Bynum
THE HISTORY OF PHYSICS
J. L. Heilbron
THE HISTORY OF POLITICAL
THOUGHT Richard Whatmore
THE HISTORY OF TIME
Leofranc Holford-Strevens
HIV AND AIDS Alan Whiteside
HOBBES Richard Tuck
HOLLYWOOD Peter Decherney
THE HOLY ROMAN EMPIRE
Joachim Whaley
HOME Michael Allen Fox
HOMER Barbara Graziosi
HORMONES Martin Luck
HORROR Darryl Jones
HUMAN ANATOMY
Leslie Klenerman
HUMAN EVOLUTION Bernard Wood
HUMAN PHYSIOLOGY
Jamie A. Davies
HUMAN RESOURCE
MANAGEMENT Adrian Wilkinson

HUMAN RIGHTS Andrew Clapham
HUMANISM Stephen Law
HUME James A. Harris
HUMOUR Noël Carroll
THE ICE AGE Jamie Woodward
IDENTITY Florian Coulmas
IDEOLOGY Michael Freeden
THE IMMUNE SYSTEM Paul Klenerman
INDIAN CINEMA Ashish Rajadhyaksha
INDIAN PHILOSOPHY Sue Hamilton
THE INDUSTRIAL REVOLUTION Robert C. Allen
INFECTIOUS DISEASE Marta L. Wayne and Benjamin M. Bolker
INFINITY Ian Stewart
INFORMATION Luciano Floridi
INNOVATION Mark Dodgson and David Gann
INTELLECTUAL PROPERTY Siva Vaidhyanathan
INTELLIGENCE Ian J. Deary
INTERNATIONAL LAW Vaughan Lowe
INTERNATIONAL MIGRATION Khalid Koser
INTERNATIONAL RELATIONS Christian Reus-Smit
INTERNATIONAL SECURITY Christopher S. Browning
INSECTS Simon Leather
IRAN Ali M. Ansari
ISLAM Malise Ruthven
ISLAMIC HISTORY Adam Silverstein
ISLAMIC LAW Mashood A. Baderin
ISOTOPES Rob Ellam
ITALIAN LITERATURE Peter Hainsworth and David Robey
HENRY JAMES Susan L. Mizruchi
JESUS Richard Bauckham
JEWISH HISTORY David N. Myers
JEWISH LITERATURE Ilan Stavans
JOURNALISM Ian Hargreaves
JAMES JOYCE Colin MacCabe
JUDAISM Norman Solomon
JUNG Anthony Stevens
KABBALAH Joseph Dan
KAFKA Ritchie Robertson
KANT Roger Scruton
KEYNES Robert Skidelsky
KIERKEGAARD Patrick Gardiner
KNOWLEDGE Jennifer Nagel
THE KORAN Michael Cook
KOREA Michael J. Seth
LAKES Warwick F. Vincent
LANDSCAPE ARCHITECTURE Ian H. Thompson
LANDSCAPES AND GEOMORPHOLOGY Andrew Goudie and Heather Viles
LANGUAGES Stephen R. Anderson
LATE ANTIQUITY Gillian Clark
LAW Raymond Wacks
THE LAWS OF THERMODYNAMICS Peter Atkins
LEADERSHIP Keith Grint
LEARNING Mark Haselgrove
LEIBNIZ Maria Rosa Antognazza
C. S. LEWIS James Como
LIBERALISM Michael Freeden
LIGHT Ian Walmsley
LINCOLN Allen C. Guelzo
LINGUISTICS Peter Matthews
LITERARY THEORY Jonathan Culler
LOCKE John Dunn
LOGIC Graham Priest
LOVE Ronald de Sousa
MARTIN LUTHER Scott H. Hendrix
MACHIAVELLI Quentin Skinner
MADNESS Andrew Scull
MAGIC Owen Davies
MAGNA CARTA Nicholas Vincent
MAGNETISM Stephen Blundell
MALTHUS Donald Winch
MAMMALS T. S. Kemp
MANAGEMENT John Hendry
NELSON MANDELA Elleke Boehmer
MAO Delia Davin
MARINE BIOLOGY Philip V. Mladenov
MARKETING Kenneth Le Meunier-FitzHugh
THE MARQUIS DE SADE John Phillips
MARTYRDOM Jolyon Mitchell
MARX Peter Singer
MATERIALS Christopher Hall
MATHEMATICAL FINANCE Mark H. A. Davis
MATHEMATICS Timothy Gowers
MATTER Geoff Cottrell

THE MAYA Matthew Restall and Amara Solari
THE MEANING OF LIFE Terry Eagleton
MEASUREMENT David Hand
MEDICAL ETHICS Michael Dunn and Tony Hope
MEDICAL LAW Charles Foster
MEDIEVAL BRITAIN John Gillingham and Ralph A. Griffiths
MEDIEVAL LITERATURE Elaine Treharne
MEDIEVAL PHILOSOPHY John Marenbon
MEMORY Jonathan K. Foster
METAPHYSICS Stephen Mumford
METHODISM William J. Abraham
THE MEXICAN REVOLUTION Alan Knight
MICROBIOLOGY Nicholas P. Money
MICROECONOMICS Avinash Dixit
MICROSCOPY Terence Allen
THE MIDDLE AGES Miri Rubin
MILITARY JUSTICE Eugene R. Fidell
MILITARY STRATEGY Antulio J. Echevarria II
JOHN STUART MILL Gregory Claeys
MINERALS David Vaughan
MIRACLES Yujin Nagasawa
MODERN ARCHITECTURE Adam Sharr
MODERN ART David Cottington
MODERN BRAZIL Anthony W. Pereira
MODERN CHINA Rana Mitter
MODERN DRAMA Kirsten E. Shepherd-Barr
MODERN FRANCE Vanessa R. Schwartz
MODERN INDIA Craig Jeffrey
MODERN IRELAND Senia Pašeta
MODERN ITALY Anna Cento Bull
MODERN JAPAN Christopher Goto-Jones
MODERN LATIN AMERICAN LITERATURE Roberto González Echevarría
MODERN WAR Richard English
MODERNISM Christopher Butler
MOLECULAR BIOLOGY Aysha Divan and Janice A. Royds
MOLECULES Philip Ball
MONASTICISM Stephen J. Davis
THE MONGOLS Morris Rossabi
MONTAIGNE William M. Hamlin
MOONS David A. Rothery
MORMONISM Richard Lyman Bushman
MOUNTAINS Martin F. Price
MUHAMMAD Jonathan A. C. Brown
MULTICULTURALISM Ali Rattansi
MULTILINGUALISM John C. Maher
MUSIC Nicholas Cook
MUSIC AND TECHNOLOGY Mark Katz
MYTH Robert A. Segal
NAPOLEON David Bell
THE NAPOLEONIC WARS Mike Rapport
NATIONALISM Steven Grosby
NATIVE AMERICAN LITERATURE Sean Teuton
NAVIGATION Jim Bennett
NAZI GERMANY Jane Caplan
NEGOTIATION Carrie Menkel-Meadow
NEOLIBERALISM Manfred B. Steger and Ravi K. Roy
NETWORKS Guido Caldarelli and Michele Catanzaro
THE NEW TESTAMENT Luke Timothy Johnson
THE NEW TESTAMENT AS LITERATURE Kyle Keefer
NEWTON Robert Iliffe
NIETZSCHE Michael Tanner
NINETEENTH-CENTURY BRITAIN Christopher Harvie and H. C. G. Matthew
THE NORMAN CONQUEST George Garnett
NORTH AMERICAN INDIANS Theda Perdue and Michael D. Green
NORTHERN IRELAND Marc Mulholland
NOTHING Frank Close
NUCLEAR PHYSICS Frank Close
NUCLEAR POWER Maxwell Irvine
NUCLEAR WEAPONS Joseph M. Siracusa
NUMBER THEORY Robin Wilson
NUMBERS Peter M. Higgins

NUTRITION David A. Bender
OBJECTIVITY Stephen Gaukroger
OCEANS Dorrik Stow
THE OLD TESTAMENT
Michael D. Coogan
THE ORCHESTRA D. Kern Holoman
ORGANIC CHEMISTRY
Graham Patrick
ORGANIZATIONS Mary Jo Hatch
ORGANIZED CRIME
Georgios A. Antonopoulos and Georgios Papanicolaou
ORTHODOX CHRISTIANITY
A. Edward Siecienski
OVID Llewelyn Morgan
PAGANISM Owen Davies
PAKISTAN Pippa Virdee
THE PALESTINIAN-ISRAELI CONFLICT Martin Bunton
PANDEMICS Christian W. McMillen
PARTICLE PHYSICS Frank Close
PAUL E. P. Sanders
PEACE Oliver P. Richmond
PENTECOSTALISM William K. Kay
PERCEPTION Brian Rogers
THE PERIODIC TABLE Eric R. Scerri
PHILOSOPHICAL METHOD
Timothy Williamson
PHILOSOPHY Edward Craig
PHILOSOPHY IN THE ISLAMIC WORLD Peter Adamson
PHILOSOPHY OF BIOLOGY
Samir Okasha
PHILOSOPHY OF LAW
Raymond Wacks
PHILOSOPHY OF MIND
Barbara Gail Montero
PHILOSOPHY OF PHYSICS
David Wallace
PHILOSOPHY OF SCIENCE
Samir Okasha
PHILOSOPHY OF RELIGION
Tim Bayne
PHOTOGRAPHY Steve Edwards
PHYSICAL CHEMISTRY Peter Atkins
PHYSICS Sidney Perkowitz
PILGRIMAGE Ian Reader
PLAGUE Paul Slack
PLANETARY SYSTEMS
Raymond T. Pierrehumbert
PLANETS David A. Rothery
PLANTS Timothy Walker
PLATE TECTONICS Peter Molnar
PLATO Julia Annas
POETRY Bernard O'Donoghue
POLITICAL PHILOSOPHY David Miller
POLITICS Kenneth Minogue
POLYGAMY Sarah M. S. Pearsall
POPULISM Cas Mudde and Cristóbal Rovira Kaltwasser
POSTCOLONIALISM Robert Young
POSTMODERNISM Christopher Butler
POSTSTRUCTURALISM
Catherine Belsey
POVERTY Philip N. Jefferson
PREHISTORY Chris Gosden
PRESOCRATIC PHILOSOPHY
Catherine Osborne
PRIVACY Raymond Wacks
PROBABILITY John Haigh
PROGRESSIVISM Walter Nugent
PROHIBITION W. J. Rorabaugh
PROJECTS Andrew Davies
PROTESTANTISM Mark A. Noll
PSYCHIATRY Tom Burns
PSYCHOANALYSIS Daniel Pick
PSYCHOLOGY Gillian Butler and Freda McManus
PSYCHOLOGY OF MUSIC
Elizabeth Hellmuth Margulis
PSYCHOPATHY Essi Viding
PSYCHOTHERAPY Tom Burns and Eva Burns-Lundgren
PUBLIC ADMINISTRATION
Stella Z. Theodoulou and Ravi K. Roy
PUBLIC HEALTH Virginia Berridge
PURITANISM Francis J. Bremer
THE QUAKERS Pink Dandelion
QUANTUM THEORY
John Polkinghorne
RACISM Ali Rattansi
RADIOACTIVITY Claudio Tuniz
RASTAFARI Ennis B. Edmonds
READING Belinda Jack
THE REAGAN REVOLUTION Gil Troy
REALITY Jan Westerhoff
RECONSTRUCTION Allen C. Guelzo
THE REFORMATION Peter Marshall
REFUGEES Gil Loescher
RELATIVITY Russell Stannard

RELIGION Thomas A. Tweed
RELIGION IN AMERICA Timothy Beal
THE RENAISSANCE Jerry Brotton
RENAISSANCE ART
Geraldine A. Johnson
RENEWABLE ENERGY Nick Jelley
REPTILES T. S. Kemp
REVOLUTIONS Jack A. Goldstone
RHETORIC Richard Toye
RISK Baruch Fischhoff and John Kadvany
RITUAL Barry Stephenson
RIVERS Nick Middleton
ROBOTICS Alan Winfield
ROCKS Jan Zalasiewicz
ROMAN BRITAIN Peter Salway
THE ROMAN EMPIRE
Christopher Kelly
THE ROMAN REPUBLIC
David M. Gwynn
ROMANTICISM Michael Ferber
ROUSSEAU Robert Wokler
RUSSELL A. C. Grayling
THE RUSSIAN ECONOMY
Richard Connolly
RUSSIAN HISTORY Geoffrey Hosking
RUSSIAN LITERATURE Catriona Kelly
THE RUSSIAN REVOLUTION
S. A. Smith
SAINTS Simon Yarrow
SAMURAI Michael Wert
SAVANNAS Peter A. Furley
SCEPTICISM Duncan Pritchard
SCHIZOPHRENIA Chris Frith and
Eve Johnstone
SCHOPENHAUER
Christopher Janaway
SCIENCE AND RELIGION
Thomas Dixon and Adam R. Shapiro
SCIENCE FICTION David Seed
THE SCIENTIFIC REVOLUTION
Lawrence M. Principe
SCOTLAND Rab Houston
SECULARISM Andrew Copson
SEXUAL SELECTION Marlene Zuk and
Leigh W. Simmons
SEXUALITY Véronique Mottier
WILLIAM SHAKESPEARE
Stanley Wells
SHAKESPEARE'S COMEDIES
Bart van Es
SHAKESPEARE'S SONNETS AND
POEMS Jonathan F. S. Post
SHAKESPEARE'S TRAGEDIES
Stanley Wells
GEORGE BERNARD SHAW
Christopher Wixson
MARY SHELLEY Charlotte Gordon
THE SHORT STORY Andrew Kahn
SIKHISM Eleanor Nesbitt
SILENT FILM Donna Kornhaber
THE SILK ROAD James A. Millward
SLANG Jonathon Green
SLEEP Steven W. Lockley and
Russell G. Foster
SMELL Matthew Cobb
ADAM SMITH Christopher J. Berry
SOCIAL AND CULTURAL
ANTHROPOLOGY
John Monaghan and Peter Just
SOCIAL PSYCHOLOGY Richard J. Crisp
SOCIAL WORK Sally Holland and
Jonathan Scourfield
SOCIALISM Michael Newman
SOCIOLINGUISTICS John Edwards
SOCIOLOGY Steve Bruce
SOCRATES C. C. W. Taylor
SOFT MATTER Tom McLeish
SOUND Mike Goldsmith
SOUTHEAST ASIA James R. Rush
THE SOVIET UNION Stephen Lovell
THE SPANISH CIVIL WAR
Helen Graham
SPANISH LITERATURE Jo Labanyi
THE SPARTANS Andrew Bayliss
SPINOZA Roger Scruton
SPIRITUALITY Philip Sheldrake
SPORT Mike Cronin
STARS Andrew King
STATISTICS David J. Hand
STEM CELLS Jonathan Slack
STOICISM Brad Inwood
STRUCTURAL ENGINEERING
David Blockley
STUART BRITAIN John Morrill
THE SUN Philip Judge
SUPERCONDUCTIVITY
Stephen Blundell
SUPERSTITION Stuart Vyse
SYMMETRY Ian Stewart
SYNAESTHESIA Julia Simner

SYNTHETIC BIOLOGY Jamie A. Davies
SYSTEMS BIOLOGY Eberhard O. Voit
TAXATION Stephen Smith
TEETH Peter S. Ungar
TELESCOPES Geoff Cottrell
TERRORISM Charles Townshend
THEATRE Marvin Carlson
THEOLOGY David F. Ford
THINKING AND REASONING Jonathan St B. T. Evans
THOUGHT Tim Bayne
TIBETAN BUDDHISM Matthew T. Kapstein
TIDES David George Bowers and Emyr Martyn Roberts
TIME Jenann Ismael
TOCQUEVILLE Harvey C. Mansfield
LEO TOLSTOY Liza Knapp
TOPOLOGY Richard Earl
TRAGEDY Adrian Poole
TRANSLATION Matthew Reynolds
THE TREATY OF VERSAILLES Michael S. Neiberg
TRIGONOMETRY Glen Van Brummelen
THE TROJAN WAR Eric H. Cline
TRUST Katherine Hawley
THE TUDORS John Guy
TWENTIETH-CENTURY BRITAIN Kenneth O. Morgan
TYPOGRAPHY Paul Luna
THE UNITED NATIONS Jussi M. Hanhimäki
UNIVERSITIES AND COLLEGES David Palfreyman and Paul Temple
THE U.S. CIVIL WAR Louis P. Masur
THE U.S. CONGRESS Donald A. Ritchie
THE U.S. CONSTITUTION David J. Bodenhamer
THE U.S. SUPREME COURT Linda Greenhouse
UTILITARIANISM Katarzyna de Lazari-Radek and Peter Singer
UTOPIANISM Lyman Tower Sargent
VETERINARY SCIENCE James Yeates
THE VIKINGS Julian D. Richards
VIOLENCE Philip Dwyer
THE VIRGIN MARY Mary Joan Winn Leith
THE VIRTUES Craig A. Boyd and Kevin Timpe
VIRUSES Dorothy H. Crawford
VOLCANOES Michael J. Branney and Jan Zalasiewicz
VOLTAIRE Nicholas Cronk
WAR AND RELIGION Jolyon Mitchell and Joshua Rey
WAR AND TECHNOLOGY Alex Roland
WATER John Finney
WAVES Mike Goldsmith
WEATHER Storm Dunlop
THE WELFARE STATE David Garland
WITCHCRAFT Malcolm Gaskill
WITTGENSTEIN A. C. Grayling
WORK Stephen Fineman
WORLD MUSIC Philip Bohlman
THE WORLD TRADE ORGANIZATION Amrita Narlikar
WORLD WAR II Gerhard L. Weinberg
WRITING AND SCRIPT Andrew Robinson
ZIONISM Michael Stanislawski
ÉMILE ZOLA Brian Nelson

Available soon:

HANNAH ARENDT Dana Villa
MICROBIOMES Angela E. Douglas
NANOTECHNOLOGY Philip Moriarty
ANSELM Thomas Williams
GÖDEL'S THEOREM A. W. Moore

For more information visit our website

www.oup.com/vsi/

Doron Swade

THE HISTORY OF COMPUTING

A Very Short Introduction

OXFORD
UNIVERSITY PRESS

Great Clarendon Street, Oxford, OX2 6DP,
United Kingdom

Oxford University Press is a department of the University of Oxford.
It furthers the University's objective of excellence in research, scholarship,
and education by publishing worldwide. Oxford is a registered trade mark of
Oxford University Press in the UK and in certain other countries

© Doron Swade 2022

The moral rights of the author have been asserted

First Edition published in 2022

All rights reserved. No part of this publication may be reproduced, stored in
a retrieval system, or transmitted, in any form or by any means, without the
prior permission in writing of Oxford University Press, or as expressly permitted
by law, by licence or under terms agreed with the appropriate reprographics
rights organization. Enquiries concerning reproduction outside the scope of the
above should be sent to the Rights Department, Oxford University Press, at the
address above

You must not circulate this work in any other form
and you must impose this same condition on any acquirer

Published in the United States of America by Oxford University Press
198 Madison Avenue, New York, NY 10016, United States of America

British Library Cataloguing in Publication Data
Data available

Library of Congress Control Number: 2022935147

ISBN 978–0–19–883175–4

Printed and bound by CPI Group (UK) Ltd, Croydon, CR0 4YY

Links to third party websites are provided by Oxford in good faith and
for information only. Oxford disclaims any responsibility for the materials
contained in any third party website referenced in this work.

Contents

Acknowledgements

To many valued colleagues, in particular Martin Campbell-Kelly, Paul Ceruzzi, Bill Aspray, and Tom Haigh, whose defining works frame the field. To Nick Linton, for his wisdom and care. To Sarah, for her patience and love. And for reading the manuscript.

List of illustrations

Chapter 1
History and computing

This book is about the history of computing. It aims not simply to chronicle the past—landmark episodes, pioneering inventors, technical innovation, and transformative technologies—but to ask how we tell the tale, and why we tell it the way we do.

Constant development and relentless innovation are features of recent times and we could argue that attempting a history is premature. It is. But there is a challenge—can history provide coherent narratives that allow us to frame the rampant and seemingly irrepressible growth of computing systems, their variety, their improbable success, and the wider context of their use?

An electronic computer is an artefact of human invention. The study of history of technology offers us precedents, models, and techniques of analysis to which a computer, as a technological object, should be amenable, or so it would seem. Marvin Minsky, a cognitive scientist and pioneer of artificial intelligence (AI), wrote in 1963:

> Man has within a single generation found himself sharing the world with a strange new species: the computers and computer-like devices. Neither history, nor philosophy, nor common sense will tell us how these machines will affect us, for they do not 'work' as did the machines of the industrial revolution... Computer-like devices

> are utterly unlike anything which science has ever considered—we still lack the tools to fully analyze, synthesize or even think about them…

This is a startling statement. Minsky asserts that what is at issue is not a difference of degree, complexity say, but a difference in kind—that computers are logically and historically novel. If computers belong to a new category of object, do they invite or demand distinct historical treatment? Minsky's claim is one of historical exceptionalism. Are the conventions of existing analysis adequate to frame the past?

Who writes history?

Much of the writing on computers, early electronic computers in particular, was not by professional historians, but by engineers, practitioners, programmers, computer scientists, pioneers, inventors, and users. Engineers tend to have deterministic models of cause and effect that do not readily transpose into models of historical process with its ambiguities and layered uncertainties. Such early accounts are, by and large, unselfconscious in that they are concerned to chronicle rather than interpret.

The late Michael Mahoney, professor of the history of science and technology at Princeton, related a salutary tale about writing history. He was a dedicated swimmer and often met up at the university swimming pool with a neurosurgeon who was a friend and colleague. Both were approaching retirement and Mike asked his friend about his plans for the future. His friend said that he had a passionate interest in the development of his field, neurosurgery, and had plans to write its history. He asked Mike about his post-retirement plans. 'I thought I would do some neurosurgery,' said Mike. The parable challenges the unexamined assumption that while no one questions the high levels of skill, expertise, experience, and knowledge required to be a brain surgeon, no equivalent expertise is seemingly required to write history.

The writings of the pioneers and early practitioners defined history of computing for several decades. To call these chroniclers amateur historians is not to derogate them. They did it for the love of it, which is the original meaning of 'amateur' when the word is disinfected from the later implications of being unqualified, unprofessional, inept, dilettantish, or even indulgent. Such contemporary accounts are for the most part not the outcome of established methods of historical analysis and often rely on unexamined assumptions about historical process, theories of technological change, and the nature of historical (as distinct from scientific) evidence.

There are professional historians who disdain chronicle histories of this kind as uninterpreted and/or 'internalist'. Yet this material provides the raw content of contemporary experience on which our histories rely. These accounts are part of the *explanandum* (that which is to be explained) of historical enterprise.

Professional historians of computing are not a populous group. The subject is not a mainstream university course. There are few full-time academics active in the field, and fewer still trained in history. There are those, mostly in computer science departments, who are indulged or tolerated by their host institutions or by renegade liberal departmental heads. Some are active researchers and offer undergraduate courses by way of a condiment, or cultural supplement, to the main diet of hard computer science. There are others who moonlight to pursue historical interests without official sanction. And there is research perched in, but not truly integrated into, the history of technology, a more established and well-populated field, and the natural parent discipline of computing history.

Museum curators are licensed and mandated to engage with history usually, though not always, focused on artefactual holdings in their collections. Curators tend to remain in post for long periods. Often this is an expression of vocational commitment which fosters continuity and depth. In traditional curatorial culture, career advancement through mobility, as in marketing and PR, is considered an offence against seriousness. Museums, as one would hope, tend to be benign hosts to historical activity.

An enduring difficulty in writing authoritative histories has been combining detailed technical knowledge of machines and systems, with wider analysis offered by business history, economic and social histories, histories of ideas, and theory. With notable exceptions historians of technology often do not have first-hand experience or specialized knowledge of the machines or of computing practice. Equally, many engineers and practitioners have little familiarity or affinity with the study of the writing of history—historiography—or of the wider study of social context.

The intellectual traditions of the two groups reflect something of the 'two-cultures' split that supposedly siloes humanities and science. Happily, a new generation of historians is emerging, bringing new dimensions to historicizing computing's past. These new analyses feature authoritative knowledge of technical detail integrated into issues of historical agency, institutional structures, personal and professional networks of influence, language usage, and human motivation.

National differences

Many who write histories are closest to the events, machines, and personalities native to their own contexts, and it is these that inevitably take centre-stage in their accounts. In the United States the machines and systems so favoured include the ENIAC, ABC (Atanasoff-Berry computer), ASCC (Harvard Mark I), Whirlwind, SAGE, UNIVAC, DEC (Digital Equipment Corporation) PDPs, IBM (International Business Machines)'s System/360, and the Cray-1 supercomputer. Most of these are universally known by their acronyms or short forms, rather than by the unwieldy variants of their expanded names (see Box 1).

Box 1 Acronymic names for some early computers

British

ACE	Automatic Computing Engine
DEUCE	Digital Electronic Computing Engine
EDSAC	Automatic Delay Storage Automatic Calculator
LEO	Lyons Electronic Office
SSEM	Small-Scale Experimental Machine
WITCH	Wolverhampton Instrument for Teaching

American

ABC	Atanasoff-Berry Computer
ASCC	Automatic Sequence Controlled Calculator
EDVAC	Electronic Discrete Variable Automatic Computer
ENIAC	Electronic Numerical Integrator and Computer
ILLIAC	Illinois Automatic Computer
JOHNNIAC	John von Neumann Numerical Integrator and Automatic Computer
MANIAC	Mathematical Analyzer, Numerator, Integrator, and Computer
ORDVAC	Ordnance Discrete Variable Automatic Computer
PDP	Programmed Data Processor
SAGE	Semi-automatic Ground Environment
SILLIAC	Sydney version of the Illinois Automatic Computer
UNIVAC	Universal Automatic Computer

Britain has a different set of canonical machines—Colossus, the Bombe, SSEM (the 'Baby'), EDSAC, ACE, Pilot ACE, LEO, Pegasus, Elliott-NRDC 401, KDF9, the Harwell WITCH. The former Soviet Union has MESM, BESM, Strela, MIR, Agat, Kronos. In Italy the Olivetti computers have pride of place, and in Germany the Z-series machines of Konrad Zuse.

Each community has its own legendary figures, protagonists, pioneers, inventors, engineers, mathematicians, villains, and heroes. Their names are respectfully and even reverentially celebrated as part of ancestral culture: Howard Aiken, John von Neumann, Presper Eckert, John Mauchly, Herman Goldstine, John Atanasoff, Grace Hopper in the US. Charles Babbage, Ada Lovelace, Alan Turing, Max Newman, Tommy Flowers, Freddie Williams, Maurice Wilkes in the UK. Sergey Lebedev, Andrei Ershov, Vladimir Melnikov, Vladimir Lukyanov in the former Soviet Union.

Few of the original pioneers of electronic computing survive, but students and junior colleagues provide first-person accounts of interactions with them and share impressions of their personalities, values, and modes of working. With first- and second-generation machines still just within living memory, and pioneers of the solid-state 'chip' era still hearty if not hale, the history of electronic computing is not yet shrouded in an inaccessibly distant past—a mixed blessing when it comes to issues of ownership of narratives, and objectivity. Recency and history are not natural bedfellows.

Names of people and machines have rich associations in their indigenous regional and national communities from which chroniclers are drawn. The machines act as cultural, corporate, and technical reference points. Figures celebrated in one community are as the unfamiliar gods of another tribe, the machines, their tribal icons. The cultural resonances in each are different.

National perspectives have fostered long-running disputes about precedence in the scramble for 'world firsts'. Which was the first electronic general-purpose digital computer, the first to feature an internal stored program, the first business computer, the first computer programmable by users rather than the computer engineers who designed and built it, and so on. Assigning distinctive precedence to all contenders helps defuse otherwise unresolved competition, appeases national pride, and ensures that fissiparous rivalries are largely friendly, superficially at least.

Emergence as a field of study

History of computing emerged as an identifiable field of study in the mid-1970s. There are two date-markers. The first is the publication, in 1973, of *The Origins of Digital Computers: Selected Papers* edited by Brian Randell, professor of computer science at Newcastle University. *Origins* is a collection of seminal papers in automatic computing from the mechanical engines of Charles Babbage in the 19th century to stored program digital electronic computers of the late 1940s. These are interleaved with linking essays by Randell.

The significance of *Origins* is both practical and symbolic. *Origins* made accessible core primary sources that were otherwise obscure and dispersed, and catalysed scholars and practitioners into historical activity. Much, though not all, of the content is written by first-generation pioneers, and the author list reads as an ancestral hall of fame. Part of what is intriguing about these contemporary accounts is that they, like diaries, were written in ignorance of what was to come. *Origins* is also a symbolic start in establishing the credentials of history of computing as a *bona fide* field of study at a time when public perception of the potential importance of computing was growing. In gathering key primary sources *Origins* signalled that computing machines had a past dating back to at least the early 19th century. *Origins* announced the field and primed its study.

A second sign of incipient maturity was the founding of the journal *Annals of the History of Computing* in 1979 by the American Federation of Information Processing Societies (AFIPS). The journal's first issue describes its aspirational scope which was 'to concentrate on the history of the computing field—covering scholarly papers and anecdotal notes, rigorously researched material and controversial remembrances, articles on the pioneers in the field and on the milieu of the time'. It had the flavour of both a scholarly journal and a club bulletin combining scholarship with news, participant accounts, profiles of pioneers, and, for many years, self-study questions and a section 'Comments, Queries, and Debate'. In time *Annals* straightened its tie and shed its folksiness. Since 1992 it has been published by the *IEEE Computer Society* in the US and it remains the flagship journal of record in the field.

For many years history of computing has been a poor relation of history of technology, itself a poor relation to the history of kings, queens, and battles. But it is far from a Cinderella activity. The Charles Babbage Institute (CBI), founded in California in 1978, is a staffed research unit at the University of Minnesota that fosters the field through research and fellowship programmes, its archive, and oral histories. Societies for information technology professionals often have active history programmes.

Master narratives

A task of history is to provide a coherent account for the gallimaufry of past events—breakthroughs, contingent occurrences, markets, users, human motivations, change, and consequence. One technique is the 'master narrative' which seeks to offer some unifying concept, proposition, or account, an organizing principle, under which otherwise unmanageable detail can be marshalled and subsumed. Without some reductionistic method of this kind we are left with a 'voluminous, unordered list of facts'. A data base without an index. Despite the youthfulness of computing history as a field of study, at least three master narratives have emerged at different times as computers and their uses proliferated.

Timeline histories

The first master narrative consists of a timeline of innovatory highlights. A typical popular account from the early 1980s starts with ancient aids for counting and remembering using knotted cords, *calculi* (small stones), and notched sticks serving as a medium of record. The tale proceeds with the transition from counting to calculation—tokens placed on counting boards, and the wire-and-bead abacus. Scaled devices then follow, marked with divisions used for calculation and measurement. Slide rules followed the introduction of logarithms in the early 1600s.

In a burst of activity in the 17th century the tale shifts to mechanical devices for calculation, mostly using numbered wheels. Then a jump-cut to the 19th century, to Charles Babbage, an English mathematician and his designs for vast automatic mechanical calculating engines. This is routinely followed by the punched-card tabulator by Herman Hollerith, used to process the otherwise overwhelming volume of data for the 1890 US census. Punched-card machines are followed in a seamless segue to IBM and the triumphs of office automation. Inserted into the office automation era is the success of manually operated desktop mechanical calculators.

Electronic computers from the 1940s then feature, followed by the invention, in turn, of the transistor, the integrated circuit (IC or 'chip') and the microprocessor. Minicomputers are covered. There is often a semi-apologetic epilogue on the personal computer boom and, in later versions of this kind, the smartphone and the internet, subjects that history struggles with while events are in wheelspin.

History framed as a linear sequence of innovatory highlights has implications for how we understand the events described. The tale so told is a technocentric one in which change is presented as innovation driven. The timeline format suggests monocausal progression in which the fact of something occurring later than something else implies causal connection with whatever came before—*post hoc ergo propter hoc* (after this, therefore because of this)—the fallacy of succession in time being equated with causal connection that students of philosophy of science are cautioned against. The notion of single causal agency implicitly endorses ideas of technological determinism and the predestiny of progress.

Need as historical cause

With the benefit of knowing how it all turned out it is tempting to present innovation as a response to pre-existing need, often critical need. John von Neumann, a stellar Hungarian-American mathematician and computer scientist, wrote a classic paper in 1945, 'First Draft of a Report on the EDVAC'. In it he described storing and running a program directly from internal computer memory. The stored program feature was near-universally adopted, and has been widely accepted as the defining feature of the modern electronic digital computer.

A variety of benefits became evident as von-Neumann-type machines were built, not all of them foreseen. Yet it is only recently that the significance of the stored program feature has been more fully explored. For all the intervening years historians, computer scientists, and others assumed that the practical and theoretical benefits that emerged were, separately or collectively, an intended response to prior need.

Framing benefits as a response to original need can be misleading. The tentative description of the internal stored program in the 1945 paper made no claim for it as the defining feature of the modern computer. The cocktail of 'needs' was projected backwards from later, and partly unanticipated, outcomes. It was the model of need being a driver of innovation that was at least partly responsible for the remarkable neglect, for over a half-century, to analyse and historicize what has been endlessly cited as the signature feature of the modern computer.

Need was deliberately invoked as advocacy in the case of Charles Babbage, who spent the major part of his life designing and attempting to construct automatic calculating engines. In the 19th-century calculations for maritime navigation, surveying, engineering, astronomy, and science relied on printed mathematical tables. It is widely accepted that Babbage's motive for his first engines was to eliminate, through mechanization, human errors in the calculation and production of printed tables which, it was argued by some, were riddled with potentially disastrous mistakes. Yet Babbage's earliest writings on mechanized calculation show that his central interest in the engines was not tabular errors but mathematics and the theoretical potential of computational machines.

Practical utility and the scourge of errors was largely retrofitted by a scientific publicist, Dionysius Lardner, in attempts to promote the engines and revive the faltering fortunes of Babbage's efforts to construct his machine. The 'tables crisis', the reality of which was vigorously and successfully opposed by experts of the day, was framed as prior need and used instrumentally to influence contemporary events.

Pre-existing need is a trope in which problem is paired with solution. Problem–solution pairing casts innovation in the role of rescuer, and a rescuer needs a damsel in distress. If there is a shortage of distressed damsels, they need to be invented. Histories often oblige.

Not all instances of need are fictitious. There are countless instances in which need is indeed a driver—Hollerith tabulators, magnetic-core memory, subroutines are some.

Seeing innovation as a response to pre-existing need eradicates any requirement for further study of historical cause, though 'cause' is a too narrowly deterministic description of the complex and nuanced agencies of historical process. Dispensing with inquiry in this way closes down accounts that better reflect more equivocal treatments, nascency for example—the birth, early evolution, and adoption of innovation. Ideas of 'cause' migrate from science, and 'need' is too strong and simplistic a proxy for historical process.

Replacement

A single line of development implies replacement, as though inventions, innovations, new techniques and devices were unconditionally welcomed in all instances by long-suffering or deprived practitioners who joyously dumped existing practice overnight in favour of the new. This was far from always the case. Uptake was sometimes hesitant, even hostile, and where acceptance was the outcome, this was not always immediate and often hard fought.

George Biddell Airy, astronomer royal, and *de facto* chief scientific advisor to government, was consulted on the potential of Babbage's automatic calculating engines. In 1842 Airy reported that Babbage's machines were 'useless'. Howard Aiken, American physicist and computer pioneer, was the conceptual driver behind the Harvard Mark I, a largely electromechanical computer built by IBM in the early 1940s. Aiken remained opposed to electronics, favouring electromechanical systems long after electronic alternatives were proving themselves. In the late 1930s Konrad Zuse, a German engineer and computer pioneer, persisted with mechanical and electromechanical solutions despite the promise of electronics. IBM itself was at first hesitant to embrace electronic technologies as transformative, preferring to make electronic upgrades to existing electromechanical inventory.

Not all beginnings were fraught. VisiCalc, a spreadsheet program conceived by a student at the Harvard Business School, Dan Bricklin, and launched in 1979 for the Apple II personal computer, was immediately embraced. It became a model for the 'Killer App', a product that quickly became indispensable to a large class of users.

Other discomforts

There are other features of historical change that do not sit comfortably with the model of relentless determinism. There is the tendency to omit long periods of dormancy during which development slows, or ideas are entirely intermitted, perhaps even forgotten, only to emerge later. Leslie Comrie, a pioneer of mechanical computation, revived, in the 1930s, the mathematical method of tabulation used by Babbage a century earlier. But by the 1930s the narrative had moved on and a hundred years is a long time for a news cycle.

Linear accounts tend to obscure simultaneous and sometimes spontaneous invention. There is no evidence of direct influence or of derivative thinking in the computers of Zuse. Working in the mid-1930s and 1940s he appears to have had no knowledge of developments elsewhere, or even of Babbage's machines. There is a temptation to deselect parallel developments and celebrate the success of those that prevail at the expense of those that do not. Chronology as history makes for compelling narratives. But its implications do less than full justice to events. Historians are alert to the pitfalls. Journalists and popularizers less so.

User-based model

If we shift from a technocentric or innovation-driven model to a user-based model the landscape opens up. Taking the human activities that the devices relieve, aid, or replace allows us to devise

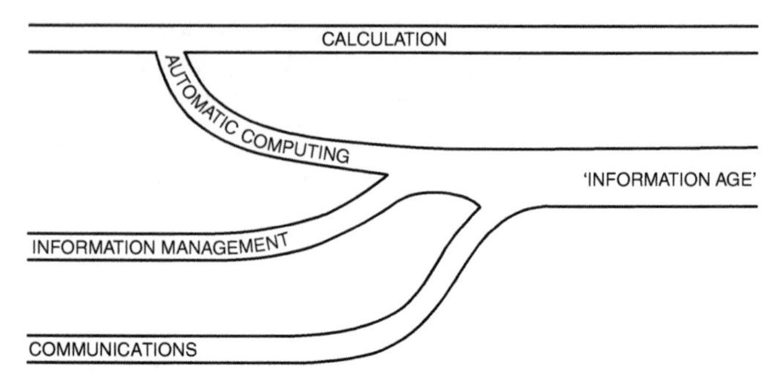

1. River Diagram showing four main threads and convergence to the 'information age'.

a more dimensioned map better able to address some of the limitations described.

The River Diagram (see Figure 1) identifies four distinct traditions or categories of human activity: calculation, automatic computing, information management, and communication, and the convergence of their related technologies into what came to be called the 'information age'.

Calculation

In the River Diagram desktop calculators are part of an unbroken thread from ancient finger-counting to the electronic pocket calculator. The defining feature of devices in this thread is that all are hand-operated and rely on the informed intervention of a human operator to deliver numerical results. Devices that populate this thread include the abacus, slide rules, graduated sectors, arithmometers, pinwheel calculators, key-driven comptometers, desktop electromechanical and electronic calculators, and, finally, the electronic pocket calculator.

There are important devices not blessed with breakthrough technology that are part of this tradition. The Comptator is a hand-operated portable adder that accumulates a running total of multiple serial additions (see Figure 2).

Typical users would have been merchants or accountants. The spacing of the digits lined up with the columns of a ledger so the device could be placed on the page and results directly transcribed. The example shown is compact, robust, beautifully crafted, and needs no batteries. Comptators were made in the US and Europe from the late 1890s and sold well in the first half of the 20th century. This period was the heyday of desktop mechanical calculators—Brunsviga, Burroughs, Felt & Tarrant, Facit are some of the famed *marques* of the time—widely used for scientific and commercial calculation, prized accessories in labs and offices.

The Comptator lives in the shadow of its celebrity cousins. It does not share the podium with contemporary 'firsts'. It is barely known. The model of innovation-driven change omits 'minor' devices of this kind, however widespread their use and practical value.

The development of manual calculators did not stop when events in automatic computing, information management, or communication made the news. It took 300 years, from the early 17th century till the early 20th century, to produce viable desktop mechanical calculators sufficiently robust and affordable for daily use. We can see the solar-powered pocket calculator as a historical epilogue to the centuries-long struggles for reliable devices for addition, subtraction, multiplication, and division—basic four-function arithmetic—reduced by the pocket calculator to the seemingly trivial. In the mature chip era, calculators, shrunk to the size of a credit card, achieved the status of promotional give-aways. Smartphones have a calculator bundled as a free app. Many computer keyboards have a key for a bundled calculator app.

The long history of manual calculation, its aids and devices, is too significant to be relegated as a backstory for the achievements of

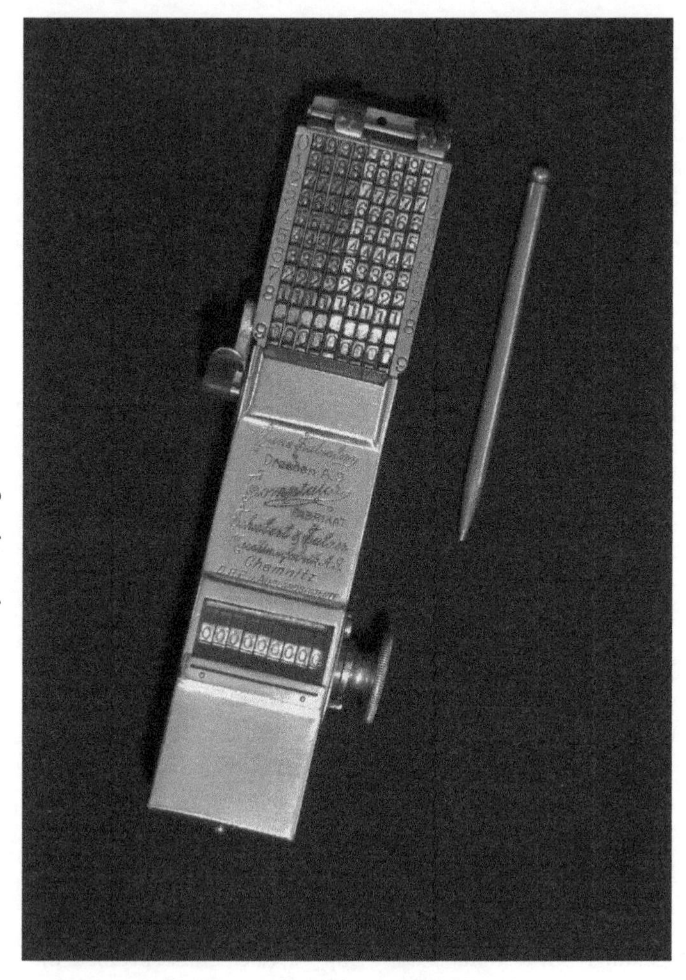

2. Comptator calculator, early 20th century, for keeping running totals.

modern computing as triumphalist accounts are wont to do. Dignifying manual calculation with a separate thread properly reinstates its traditions and practices, reminds us of the unbroken continuity of need and use, and rescues it from a supporting role that does less than full justice to the richness of its history.

Automatic computation

Automatic computing is a separate thread that starts in earnest in 1821, with Charles Babbage and his designs for vast automatic mechanical calculating engines. Babbage failed to complete any of his engines in their entirety. With the death, in 1852, of Ada Lovelace, who collaborated with Babbage, and of Babbage himself in 1871, the movement to automate calculation lost its two most vigorous advocates. There is a gap of about a century, until the 1930s and 1940s, when automatic computation revived and then boomed. Having a separate and unbroken thread, despite the developmental gap, reflects the uninterrupted aspirations and ambitions of automatic computation, and is truer to historical process than a model populated by unrelieved innovation.

Information management

The information management thread represents the activities and techniques of information processing, storage, and retrieval. Like the first thread, its roots can be found in manual methods of antiquity. Tokens used as counters to record quantity of goods were used earlier than at least 3000 BC. Knotted cords recorded quantities and the dates of important events in a retrievable way. The representational power of writing was transformative. Characters, pictograms impressed on clay tablets, inscribed on papyrus, vellum, and paper became the dominant medium for organization, record, and dissemination. 'There exists no memory except upon paper' wrote Airy in 1854, a near-obsessive devotee of a needle-and-hole filing system that he devised in 1837 for document archiving and retrieval.

Pre-mechanized information processing was through human agency—statisticians, clerks, transcribers, typesetters, collators. The growth of the machinery of state during the process of industrialization, of government, business and of economic management, was accompanied by mechanization. The practical and symbolic event in this thread is the introduction of punched-card machines by Herman Hollerith for the 1890 US census. The focus of this thread is high data volumes and throughput rather than numerical computation.

Communication

The final thread is communication. Gestures and speech are properly candidates for inclusion, though the focus here is communication over distance and the related technologies. Telecommunication is nowadays associated with electronic communication though this ignores pre-electronic and pre-electrical histories. The prefix 'tele-' is Greek for far off, so, strictly, telecommunications refers to any means of information transmission or exchange at a distance. Ancient precedents include Greek fire-beacons, often in a relay chain, 'talking drums' used for subtle and sophisticated communication in pre-literate sub-Saharan Africa, military bugles, flags, semaphore, occulted lights, and postal systems. Electrical and electronic devices align more closely to present-day perceptions of communications—the electric telegraph, telephone, fax machine, radio, teletext, television. More recently data-routing over networks of fibre-optic cables, line-of-sight microwave networks, and geostationary satellites are signature features of this thread.

Deconstructing the first master narrative

The River Diagram unplaits the braid of the linear timeline narrative into four separate threads. If we now look at the chapter sequence of our first master narrative we can see that it is constructed by splicing together segments lifted from different developmental threads and butted together to create a serial story. When automatic computation went dormant in the latter part of the 19th century, Hollerith and punched-card tabulators took up the tale. A linear timeline might suggest influential connection. There is little, if any, evidence that Hollerith's machines owe anything to Babbage, notwithstanding the use of punched cards by both. The computational capabilities of Hollerith's machines were relatively trivial (cumulative addition). The large volume of census data was an information management problem rather than a computational one. The two belong to different traditions and principles of practice.

Date markers

There is time progression in the River Diagram starting with distant antiquity on the left and the modern era on the right. The branching from calculation to automatic computing dates from 1821, with Babbage's first attempts to mechanize mathematics. The confluence of information management with automatic computing can be dated from 1913 with the installation, in New Zealand, of the first Totalisator—vast electromechanical multi-user systems for managing dog and horse track betting. Totalisators accumulate bets in real time and compute the winning odds for live public display. Unlike Hollerith's census machines, in the Tote computational capability and information handling are more equally balanced. Finally, the convergence of communications with the main trunk can be dated to 1970 and early automatic digital telephone exchanges.

Other master narratives

Timeline models, especially those with a single developmental line, are widely discounted by professional historians. I have used them here as a device, a straw man, to talk about historiography, and to illustrate how some narratives serve history better than others. There is a further reason. In my experience, chronology remains the most prevalent model of history among those who are not professional historians. So it seems a good place to start. Excavating its limitations allows us to judge other narratives that have emerged, treatments that benefit from ideas migrated from the humanities—relativism, agency, audience, social, cultural, and economic context.

A second master narrative frames the electronic computer as a scientific tool, a machine for mathematical, scientific, and engineering calculation. Its reference points are the machines created between the late 1930s and early 1950s when new principles, techniques, and terminology were developed and contested. The timeline model follows a thesis of unfolding innovation, a progress narrative. The second master narrative frames the computer in terms of what it was used for, a unifying concept of operational utility.

A third master narrative is that of computer as 'information machine'. This broadens the reach to business administration, information management, and corporate office systems. Its scope extends to real-time operation and software for non-mathematical purposes—text editing, communications, and data management. The narrative embraces the era of large centralized machines (mainframes), the personal computer, email, the dated futurism of the 'information superhighway', the internet, and the web.

How do these narratives serve our most recent past—networking, gaming, social media, leisure, and entertainment: defining features of the past twenty years?

Chapter 2
Calculation

Calculation involves manipulating numbers according to rules. Numbers are abstract. We do not experience or manipulate them directly. What we do instead is manipulate representations of number in physical form. We can represent the value of a number by the position of tokens on a counting board, graduated markings on a scale, the positions of beads in an abacus, by how far a gear wheel is turned or a disc rotated, or, in electronic systems, by voltages. For a machine to compute requires the physicalization of number in some form so that number can be acted on in accordance with arithmetical, mathematical, or logical rules.

Number need not always represent quantity. Numbers can represent letters of the alphabet, identify the spare part for a car, a prisoner, the classification of a book. Here numbers act as identifiers or numerical labels untethered, at least partly, from quantity.

More expansive histories of computing look to the origins of computing in early number systems—Egyptian, Chinese, Roman, Greek, Hindu-Arabic (our current European system), and so on—and how these are used to quantify and record. So one way of structuring pre-electronic histories of calculation is through the ways in which number has been physicalized (representation), the ways in which representations are manipulated (mechanization), and the extent to which human agency, mental and physical, is transferred to machine (automation).

Physical aids

There is a profusion of physical aids and devices used through the ages to record information, as a proxy for human memory, and to aid and augment our ability to calculate.

The use of knotted cords in China 'for the administration of affairs' is thought to date from as early as 2800 BC. The knotted tassels of a present-day undergarment worn by religious Jews are ritual reminders of the commandments, a legacy from biblical times. Roman tax collectors used knotted cords to record liabilities and receipts as recently as 300 AD. Inca tribes used knotted cords as a mnemonic for the timing of festivals, poems, and the names of rulers. Their use is not part of a mistily distant past. In India illiterate recorders for the 1872 census used cords with different colours for man, woman, girl, and boy, this less than twenty years before Hollerith devised his mechanical punched-card tabulators for the 1890 US census.

Knotted cords belong to two threads in our scheme. Used for arithmetic, they are part of the calculation thread. Used as a medium of record, they are part of information management. Calculation entails process. In the case of knotted cords, whatever processing is involved—coding a number into a pattern of knots—resides in the human agent, not in the device.

Abacus

The abacus was originally described as a layer of fine dust or sand shed on a flat surface onto which the user inscribed lines and marks on which pebbles, counters, or tokens were placed and moved. The abacus uses a positional system of value—values are indicated by the placement of a token in relation to markings on the table. A token on one line typically represents a fixed multiple of a token on the line below—tens, and units, for example.

An abacist would cast counters on the table and move them around to perform addition, subtraction, multiplication, and division. In France counters were called *jetons* from *jeter* to throw, and *jeton* meaning a token is still used in France. The user directs and carries out the calculation and it is only through moving the tokens, through physical intervention, that calculation can proceed. Manipulation is carried out by the user who abides by the necessary conventions and procedural rules. The markings on the table are no more than an aid.

Tokens can be placed freely on the table. Their movements are not constrained by any feature of the apparatus but only by the rules necessary to achieve meaningful results. Tokens can be placed in error and produce an incorrect result. They can also be adrift of logically valid markings and show an indeterminate result that has no meaning. Unwanted disturbances that derange information or devices are, in modern information theory, called 'noise'. Without physical restriction on their placement, tokens are subject to derangement and we would now say that the table abacus has low noise immunity.

Mechanism

The wire-and-bead abacus has beads threaded on wires held in a frame. Here we find incipient mechanism—motion under mechanical constraint. The wires physically confine the beads to one degree of freedom—sliding along the wires. Moved to the ends of the wires beads come up against the frame and are automatically aligned with beads at the extremities of other wires. Beads cannot occupy positions in the spaces between the wires, a logical and physical no-man's land.

For the wire-and-bead abacus, as with the table abacus, a calculation can only proceed through the directed agency of the abacist. The algorithm—the procedural rules to achieve a valid result—resides in the human agent, not in the machine.

There are many variants of wire-and-bead abacuses—the Chinese *suanpan*, the Japanese *soroban*, and the Russian *schoty* are the three best known. They differ in configuration—number of beads, number of wires, and the physical grouping of beads in the frame. The feature of partial physical constraint of the moveable beads applies to all. Constraining the travel of the beads reduces susceptibility to derangement and so improves noise immunity.

The Sumerian abacus appeared between 2700 and 2300 BC, and abacuses in some form have been with us for a long time. In Moscow in the mid-1990s, I witnessed a customer in a bread shop disputing the amount of change she was given, protesting that she did not trust the result from the electric till. It was only when the shopkeeper did the same calculation using a *schoty*, in her presence, that the customer was placated.

The historian E. P. Thomson warned against 'the enormous condescension of posterity' in attitudes to superseded trades and practices. It would be wrong to trivialize the powers and sophistication of the abacus. Michael Williams, historian of computing, reports that in 1947 a Japanese *soroban* in the hands of a highly skilled abacist went head to head with the most up-to-date electrically driven four-function calculator in a contest of speed and accuracy. The tests involved a mix of numerical tasks including addition, subtraction, cumulative summation, and multiplication. In four of the five tests, the abacist won. The electrically driven mechanical calculator prevailed in the multi-digit multiplication tasks, but only just.

Scaled devices

The 17th century saw a surge in devices for calculation. One group of these are scaled devices where the values of various mathematical functions are marked as graduations on a scale—logarithms, trigonometric functions, squares, cubes, reciprocals. The quadrant, proportional compass, sector, and slide rule are in this class.

Taking the quadrant as an example will help us identify the defining features of how we might classify these abundant devices. A quadrant is so called because it is in the form of a quarter circle. They were used extensively throughout the 17th century. One, by William Leybourn, an English mathematics teacher and surveyor, which dates from the late 17th century, is shown in Figure 3. It has scales on the front and back faces that include trigonometric functions, and tables for the geometry of circles.

There are scales for multiplication, division, the computation of squares, cubes, roots, astronomical observations for twelve of the principal stars, the daily position of the sun for each day of the year, zodiacal signs, and astrological aids. There are also scales for the design of sundials. Leybourn's 260-page book describing the device does not exhaust its uses.

The point of this litany of features is to ask, in what sense might these devices be said to compute? They are part of the canon of extended histories of computing, but do they belong there at all?

The astronomical data are empirical, based as they are on observation, and values spread along a scale are read off by eye. In the scales for mathematical functions the spacing of the graduations is pre-computed before manufacture with the tapered intervals determined by the function the value of which is being sought. These calculations are carried out by the designer.

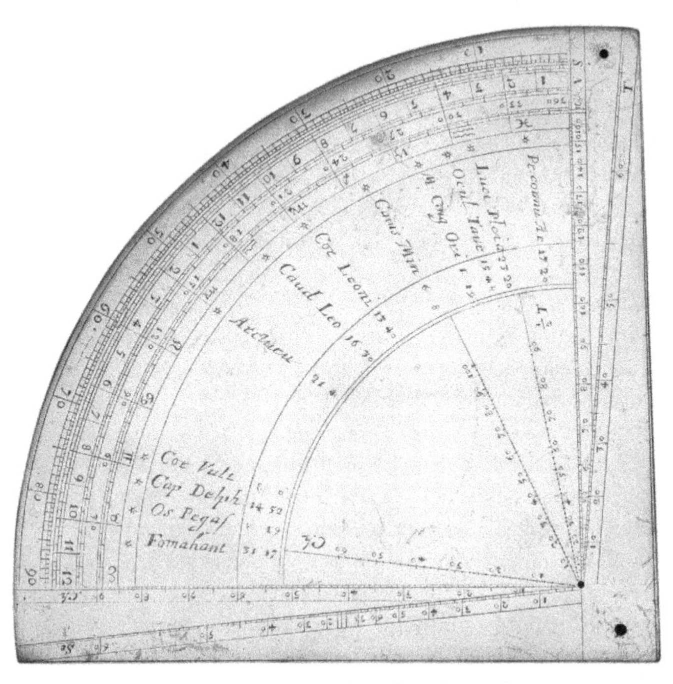

3. Quadrant by William Leybourn, front face, late 17th century.

Graduations are bunched in some part of the scales and less crowded in others. The user performs a mathematical calculation by a manual procedure. Taking a pair of dividers, the user takes a measurement on one scale, lays it off against another, and reads off a result. Whether the process uses empirical or mathematically derived data for the scales, in neither case does the device compute. The process is essentially one of measurement.

The markings on a scaled device are a way of organizing numerical information, and using it, in effect, as an analogue look-up table. So, scaled devices belong to the information management thread as well as the calculation thread. Before such devices were abducted by computing history they were classed as mathematical or astronomical instruments, and for the most part remain so classed in museum collections.

Slide rules

The most evolved device in the class of analogue scaled devices is the slide rule, analogue in the sense that number values are represented as a continuum with values between graduations provided by visual interpolation, that is to say, by subjective judgement.

John Napier, a Scottish mathematician, published, in 1614, a description of logarithms. An immediate advantage of logarithms is that the product of two numbers can be found by adding their logarithms; and division, by subtraction. This property reduces multiplication and division, formidably difficult to implement mechanically, to simple addition or subtraction.

In 1620, Edmund Gunter, an English mathematician and astronomer, plotted a logarithmic scale on a 2-foot piece of wood and added two logarithmic values by adding lengths using a pair of dividers in order to multiply two numbers. The slide rule followed soon after by placing two Gunter scales alongside each other and sliding them so as to allow the addition or subtraction of logarithms without the need for dividers.

As with the quadrant, the slide rule does not itself compute. Its mathematical scales are pre-computed to give the spacing of the divisions. To perform a calculation, the user slides one of the scales along another and moves a cursor to read off a result. The process is essentially one of measurement of length with the algorithmic procedure provided by the user.

Exotic variants of slide rules for special purposes proliferated and give insight into the social history of the times. Examples from the late 19th century include scales for calculating the weight of cattle, the volume of timber, estimating varieties of interest rates, evaluating excise duties with conversion scales for cubic inches to bushels, finding the mean diameter of a cask, and for a host of specialized engineering applications.

Slide rules were relatively inexpensive. But accuracy was limited. Reading graduated scales by eye requires an element of judgement, especially for the last decimal places. Precision was variable, poorer when interpolating divisions in the crowded sections of the scales. Accuracy was typically limited to between two and four figures with increasing uncertainty in the last figures.

In attempts to improve accuracy scales were extended using circular, cylindrical, and helical forms where scales were wrapped around the stock or barrel. A cylindrical rule by George Fuller in 1878 has a logarithmic scale wound round the surface fifty times (as in a screw thread), and gave a working length of just under 42 feet giving reliable results to four places. A telescoping Fuller's rule featured a logarithmic scale 83 feet long and could be read to four and sometimes five figures, the limit of precision in a scaled rule that was still portable.

When it came to a comparison of precision between slide rules and printed tables it was no contest. Leslie Comrie, one-time superintendent of the Nautical Almanac Office, wrote in 1948 that 'today schools are equipped with 4-figure tables, which are ten times as accurate as the common 10-inch slide rule with which the great majority of engineering calculations are done'.

Slide rules as a class had a lifespan of some 350 years from invention in the early 17th century to their final and near-complete demise in the early- to mid-1970s, when electronic calculators rendered them obsolete seemingly overnight. For well over a hundred years, from the mid-19th century onwards, slide rules were the mainstay of engineering and scientific calculation where two- or three-digit accuracy was sufficient. During the

1960s one of the largest manufacturers of slide rules in the US, Keuffel and Esser, produced 60,000 rules per year. Well into the 1970s engineering students could be readily identified on campus by the presence of their slide rules, an insignia of their profession, much as a stethoscope is for a doctor.

Precision

The levels of precision offered by scaled devices were adequate for many applications in engineering and science given that practical measurement was typically limited to two or three decimal places anyway. But interest payments, assurance premiums, annuities, and general accounting required exactness to the level of pennies in thousands, and hundreds of thousands of pounds. Land surveys, especially for cadastral tables used for property taxation, required calculations with the precision of metres in distances of kilometres and tens of kilometres. These levels of precision were unattainable using analogue scaled devices, and calculations involving more than three or four decimal places relied on printed tables pre-calculated to four, seven, ten, twenty, and sometimes more than thirty decimal places.

Abstracting computational process as something distinct, and defining it as mechanical manipulation of number, is, it must be said, anachronistic when applied to early scaled devices. Doing so is a backwards projection from the later, near-universal, adoption of digital technologies, early practical realizations of which are found in mechanical calculators. Modern histories selectively confer special significance on devices and techniques from the past seen as precedents for the digital technologies that later prevailed. Isolating elements of what is seen as distinctive, or even defining, in the light of how things turned out has the whiff of Whig history, that is, framing the past in terms of precedent for known or favoured outcomes, a practice frowned upon by many unforgiving historians. I thought it best to confess now.

Mechanical calculators

The realization of workable mechanical calculators turned out to be more difficult than expected. It took some 350 years to find an effective solution.

The earliest device with identifiable mechanisms for arithmetical calculation is the Calculating Clock by a little-known polymathic German professor, Wilhelm Schickard. The original device, constructed in 1623, is lost without trace, and a half-finished second machine made for Johannes Kepler, mathematician and astronomer, was lost in a fire in 1624. Schickard's calculator was unknown until 1935, when a scholar found a rough sketch in Kepler's papers and, in 1956, a second drawing and brief description. These findings were not revealed until 1957. Schickard's Clock is now part of the canon but does not feature in historical accounts written before its discovery was made public.

Schickard's device implemented, mechanically, the arithmetical procedures of addition and subtraction. The procedural rules, for at least part of the multiplication process, resided for the first time in wheelwork rather than in the mind of the user. Details of Schickard's calculator, unknown to philosopher-designers or to later historians for over 300 years, had no known influence on what followed.

Before details of Schickard's 'clock' came to light the two most celebrated 17th-century devices that sought to mechanize four-function arithmetic were the Pascaline by Blaise Pascal, French mathematician, physicist, and inventor, and the 'stepped reckoner' by Gottfried Leibniz, German polymathic philosopher and mathematician. The significance of the Pascaline and the Stepped Reckoner is as much to do with what they symbolized as with their capabilities. In practical terms, the Pascaline was only partly successful, and it remains questionable whether Leibniz's worked effectively at all. Both are amongst the exalted artefacts in the mechanical prehistory of computing. By impersonating limited aspects of human reason these aspirational machines symbolize the first faltering steps in the mechanization of mind-like activity.

The Pascaline

In 1642, aged 19, Pascal attempted the design of his first machine. His father, Etienne, was a commissioner of taxes and the stimulus appears to have been to relieve the effort and tedium of adding up interminable columns of figures for tax levies with which Blaise appears to have helped. The main technical accomplishment of the Pascaline was a delicate device for the carriage of tens, a problem that bedevilled calculator design for generations and that found no reliable practical solution until the middle of the 19th century.

Schickard's method for carrying tens was simple. Each time a number wheel went one full rotation, a single tooth on that wheel incremented the next higher wheel by 1. A problem arises when several carries are needed in the same addition. The worst case of this is when say a 1 is added to a row of 9s, i.e. 1 added to say 99999. Each of the wheels at 9 needs to increment by 1 and nudge the next higher wheel up. So the first number wheel needs to drive all the rest. The problem was that the materials from which the mechanisms were made—wood, ivory, and soft metals—were simply not strong enough for a cascade of more than about five digits. Pascal's mechanism offered a solution. He claimed that his device allows 'one thousand or ten thousand dials' to be moved with the same ease as one, and that the machine frees the operator from the 'vexation' of carrying and borrowing tens using pen and paper. What was muted in his claim was that the mechanism was not entirely reliable and prevented turning the dials in reverse as required for subtraction.

Analogue and digital

There is a second feature of the Pascaline that was more easily solved and was a recognition of a feature of computational design that endures to modern times. The problem was how to physically inhibit intermediate number values. What is at issue here is the distinction between 'analogue' and 'digital', though it must be emphasized that these terms were not used by the savants of old. The meanings of these terms were not sharpened until the 1940s.

A gear wheel, or circular dial, turning on a shaft is inherently analogue in the sense that it can take up any rotational position when turned. Any and all positions in the rotational continuum are physically viable. In desktop calculators each digit in a number has its own wheel. So the value 2.5 is not represented by a single wheel half-way between 2 and 3 but by two wheels, one wheel at 2 and the next adjacent wheel at 5. This is a digital form of representation in which only full integer numbers are logically valid. A wheel positioned imprecisely, somewhere between two numbers, as it might be when entering a number by hand using a stylus to turn the wheel, is physically viable but logically indeterminate.

So a design requirement was to ensure that the motion of the wheel is discretized by some physical action to inhibit positions of the dial between whole numbers. The difficulty was readily solved, at least partially, by a sprung lever pressing a roller, for example, into slots or indentations in the number wheel. This biases the wheel to occupy only the preferred positions that correspond, in a decimal system, to the numbers 0 to 9. The use of indents and sprung levers reduces susceptibility to derangement and so, as we have seen, increases noise immunity.

A machine that thinks

After his first five-digit calculator Pascal produced extended versions with six, eight, and ten digits. He records that he made as many as fifty models in the course of development, and tested the machine's portability, robustness, and immunity to derangement by riding with them over rough terrain. For all his fastidious efforts, reliability was elusive.

Pascal's calculators were paraded before royalty and demonstrated in the drawing rooms of merchants, government officials, aristocrats, and university professors. Most of the devices were ornate curiosities, philosophical novelties insufficiently robust for daily use, and often only partially working. They were expensive and not many were made. For all the ingenuity of their makers and their seriousness of purpose, mechanical calculators prior to the 19th century were largely *objets de salon*, many exquisite and delicate, sumptuous testaments to the instrument-maker's art, but unsuited to daily use in trade, finance, commerce, science, or engineering.

The Pascaline was strictly an adding machine. It was a practical albeit imperfect example of transferring to a machine something that up to that point in time was achievable only by mental activity. If doing arithmetic, which required thinking, could be reduced to mechanical procedure, what were the implications for human reason and its supposed uniqueness to our species? If rule-based machines were capable of mind-like behaviour then what of human creativity?

Computers and AI feature prominently in modern debates about mind, matter, thought, and consciousness. Matthew Jones, a historian of science and technology, takes care to point out that early calculating machines barely feature in 18th-century discussions about reason and matter. There are clearly dangers in projecting backwards the priorities of our own intellectual landscape onto earlier times. Debates about mind and machines were re-invigorated in the 19th century by the prospect of fully automatic calculating engines that required no informed human intervention other than motive force. 'Teaching wheelwork to think' had starker implications in the 19th century, given the widening fault-lines between science and religion as rivals for ownership of a world view.

Leibniz's calculator

The second calculator, celebrated in the early history of mechanical calculators, is that of Gottfried Wilhelm Leibniz, a German polymath famed for his formidable achievements in mathematics, philosophy, logic, and theology. He is perhaps now best known for the calculus for which he shares priority with Isaac Newton. The story of his efforts is as much a technical tale as it is a salutary study of how we construct historical narratives.

Leibniz had begun to consider a calculating machine around 1670 when in his mid-20s, prompted, he recalls, by a pedometer that recorded the number of steps taken when walking or running. He examined the Pascaline in Paris in 1672, made notes on its working, and set about extending Pascal's adding machine to a four-function calculator capable of multiplication, division, addition, and subtraction. From 1672 until his death in 1716 Leibniz attempted at least three versions of his calculator. By all contemporary accounts none worked reliably.

Standard histories of computing credit Leibniz with the invention of what is known as the 'stepped wheel', 'stepped drum', or 'Leibniz wheel' that was a key part of the multiplication mechanism of his machine (see Figure 4).

The stepped drum is routinely identified as the defining feature of calculator design for the next 200 years, until the 1870s, and well-established accounts state explicitly, or strongly imply, that the significant machines that followed the Leibniz calculator were based on Leibniz's wheel.

Others attempted calculating machines: Sir Samuel Morland and Robert Hooke in the 17th century; Giovanni Poleni, Philipp

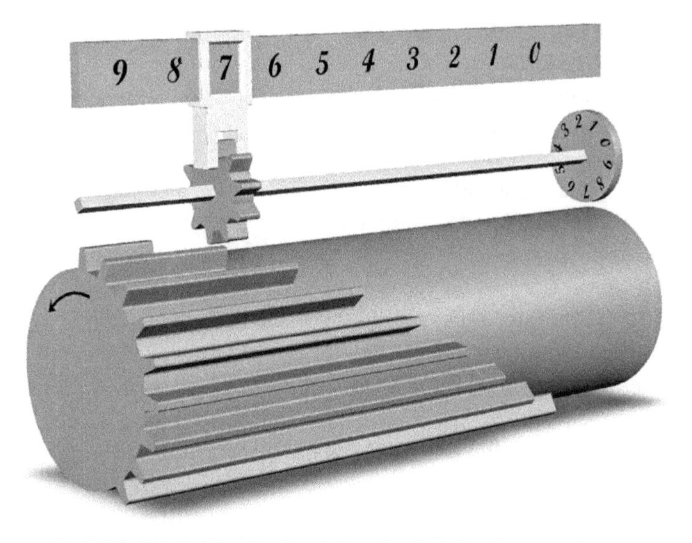

4. The 'Leibniz wheel' or 'stepped drum', a defining feature of mechanical calculator design for 200 years.

Matthaus Hahn, Charles Mahon (3rd Earl of Stanhope), and Johannes Müller amongst others in the 18th century; and, famously, Thomas de Colmar in the 19th century. Many of these machines worked and, in the rich tale of these efforts, historical accounts privilege the machines by Hahn, Stanhope, Müller, and de Colmar. They all used a stepped drum, and part of their credentials for a place on the podium is their supposed indebtedness to the Leibniz wheel, with their work framed as a vindication of Leibniz's great invention.

What is puzzling about this narrative is that the defining technical details of Leibniz's machine did not emerge until the late 1890s, over a hundred years after machines by Stanhope, Müller, and Hahn, and nearly eighty years after Thomas de Colmar announced his arithmometer. When we examine how these later makers might have known about Leibniz's mechanism the puzzle deepens.

Leibniz was notoriously secretive about technical detail and his publication in 1710 reveals nothing specific about the internal workings of his machine. He consistently exaggerated the state of completion and capabilities of his machines, at the same time concealing details of the mechanisms themselves. His priority for the stepped drum for use in a calculator is widely accepted: his sketch, from around 1673, of the multiplying mechanism explicitly shows the stepped drum described as a 'wheel with unequal teeth'. But transmission by publication is problematic. It was not until a report in the late 1890s (unpublished) that technical details were revealed. Until then it seems that the stepped drum was not even commonly identified as Leibniz's invention.

Another possible line of transmission would have been through access to a physical model. This too is problematic. The one surviving model was sent to Göttingen in 1764 but few knew of its whereabouts. Correspondence indicates that it was still there in 1799 and was then seemingly lost until the late 1870s when it was rediscovered, apparently in an attic.

Re-inventing Leibniz

It seems from this chequered history that lack of disclosure, and the absence of a machine to examine, meant that later calculator builders most likely did not know about the Leibniz wheel. How then did it come to be that all stepped-drum calculators that followed are said to be based on Leibniz's wheel? How has history transformed Leibniz, the inventor of a machine that is not known to have worked, whose essential contribution remained unattributed for over a century, into the venerated ancestor of the mechanical calculator?

Leibniz's rehabilitation began in the late 1870s when the one surviving model was found, in Göttingen. (Leibniz died in 1716.) A report on it in 1880 repeated Leibniz's own bloated claims for his wondrous machine and concluded, without evidence of influence, that the structural similarities between Thomas de Colmar's arithmometer and Leibniz's machine were sufficient to establish that Thomas must have known about the Leibniz machine, and that Leibniz was 'the originator of all later mechanical calculators using the stepped-drum principle'.

This delighted German companies manufacturing versions of the increasingly successful French arithmometer by Thomas. Politics may well have played a role. Germany had summarily defeated France in the Franco-Prussian war that ended in 1871. Assigning priority to Leibniz for the arithmometer's signature mechanism allowed German manufacturers to frame the machine of the vanquished French as imitative. And it seems that it is this confected narrative, with its source in commercial rivalry and national pride, that has been uncritically consolidated into the established account.

The idea of the lone genius resonated with contemporary cultural values. At the time of Leibniz's rehabilitation, the Romantic movement was nearing its peak, and celebrating Leibniz as 'heroic inventor' is a trope of the period in which individualistic inspiration and personal creativity were prized. Where, in a technocentric timeline history, might we find a place for Romanticism and the Franco-Prussian war?

Turning point

A turning point in the tale is the arithmometer and the transition of the desktop mechanical calculator from aspirational curiosity, with craft origins, into an industrial product. Thomas de Colmar, director of an insurance company in Paris, announced his arithmometer in 1820—a four-function calculator for multiplication, division, addition, and subtraction. It featured sliders for entering numbers, numbered dials to display results, and a moveable carriage that could be lifted and shifted left or right, not unlike Leibniz's calculator in externals.

The arithmometer is routinely described as the first successful commercial calculator. But it was far from the instant success that this sweeping strapline implies. The first model in 1821 was some way from a device ready for market, and it took over fifty years of modification and improvement before arithmometers attracted even modest sales. Technical niggles dogged the device. But in time it sold moderately well. Success was hard won. Thomas's role was that of an entrepreneur and improver rather than that of the savant designers of the 17th and 18th centuries.

Insurance companies adapted their procedures to exploit the arithmometer and in turn became dependent on them. By 1872 institutional and professional users included banks, large financial institutions, insurance companies, actuaries, government departments, observatories, laboratories, universities, engineers, and surveyors. The arithmometer rode the wave of the burgeoning industrial movement and the seemingly insatiable demand for calculation in science and commerce. But by the 1890s a new generation of machines, compact, lighter, and finally reliable, began to rival the solidly worthy (though bulky) arithmometer.

Pinwheels and keyboards

The desktop calculator became a mass-production, commercial product on the back of two devices. First is the pinwheel, conceived in 1872 by an American, Frank Baldwin, and patented in 1875. The pinwheel for the most part replaced the arithmometer's stepped drum in calculator design. Pinwheels are lighter and smaller than the standard stepped drum, the irreducible bulk of which limited how sprightly an arithmometer could be made. It is also easier to set up numbers using the pinwheel levers than using sliders as for the stepped drum.

Not long after Baldwin patented his pinwheel, a Swede, Willgodt Odhner, working in Russia, produced a similar device but laid no claim for priority. Large-scale manufacture of Odhner machines started in 1886 in St. Petersburg. Baldwin's first machine appeared on the US market in 1875. Production of developed versions ramped up after 1912 with the founding of the Munroe Calculating Machine Company.

The second feature that ushered the desktop calculator into wider acceptance was the use of typewriter-like keys to enter data. Entering a digit on a Baldwin-Odhner pinwheel calculator required moving a small lever, the end of which protruded outside the casing, to the position for the value of that digit. Multi-digit numbers were entered one digit at a time. Entering numbers in this way requires fine muscle control, care, and concentration. Once the numbers were entered the operator turned a crank handle that drove the mechanism to complete the calculation. In contrast, striking a key entered the digit value and the same action drove the calculating mechanism to perform the addition. The user is spared the extra operation of cranking a handle. Striking a key is also more tolerant of brusque action than is setting a lever. And it is quicker.

Dorr E. Felt, a machinist in the Pullman Company in Chicago, made a prototype of a key-driven calculator, a comptometer, on his kitchen table in 1884 using rubber bands, a macaroni box, and meat skewers. The tale of Felt's breadboard attempt, with Felt in the role of heroic inventor, has the flavour of myth, but the episode is credibly evidenced. He developed an engineered version and added a listing device that printed number-entries and results.

Printout and the speed advantages of keyed operation had great attraction for bankers, accountants, bookkeepers, statisticians, merchants, and auditors. Commercial environments, where speed was at a premium, came to favour key-driven comptometers over the Baldwin-Odhner pinwheel machines. Calculator manufacturers benefited from mass-production techniques using standardized interchangeable parts, a practice established in the typewriter and firearms industries. Felt joined forces with Robert

Tarrant, a Chicago businessman, to create the Felt & Tarrant Manufacturing Co., which began production of key-driven comptometers in 1889.

The office appliance industry

Mechanical calculators, alongside typewriters, became commonplace in offices, laboratories, banks, clearing houses, universities, and government departments. Production, marketing, and distribution created a burgeoning and lucrative office appliance industry and corporations vied with each other for market dominance. Armies of clerks and operators could quickly produce responsive management data, and this fuelled the growing bureaucratization of commerce, industry, and labour.

There is no comprehensive business history of the mechanical calculator industry between 1890 and 1930, the period of major growth and consolidation. But surviving company records, and social histories, evidence the transformative role of these machines in commerce, finance, industry, and trade. The big players traded alongside countless smaller suppliers in a crowded market. In the US, Burroughs and Felt & Tarrant, two of the largest, were energetic rivals for market share. The brands, Brunsviga in Germany and Facit in Sweden competed against the US in European markets. In 1909, a boom year, Burroughs sold 15,763 machines. By 1912, 20,000 Brunsvigas were sold, mostly in Europe.

Well-established markets were receptive to the innovations that followed. Powered comptometers, driven by electric motors, appeared in the 1930s. Machines using electromechanical relays, then electronic valves (called vacuum tubes in the US) followed. Then came transistorized machines. Finally, ICs, liquid crystal displays, and solar cells in place of batteries featured in the latest and possibly the last generation of calculators in this thread.

The calculator had come of age. What had been a curiosity had become a commonplace commodity. Portable, rapid, reliable, and power-efficient calculation ceased to be a problem some 350 years after Wilhelm Schickard's first faltering attempts in the 1620s.

The replacement thesis

The concept of a thread is a linear one and we should not be misled into thinking that each new device replaced the old. The tenure of each was finite and the differing lifespans of production and use overlapped, as would fibres in a natural thread. Rates of take-up, and the taper into disuse, were sometimes gradual, sometimes abrupt. Use, need (perceived and real), production, expectation, and opportunity were entwined.

Figure 5 shows some of the overlaps in product life and use since 1800. Slide rules in various forms were used for over 300 years. Their demise was sudden and complete with the advent of hand-held electronic scientific calculators in the 1970s. Pinwheel machines, promoted in the 1880s, had a long tail, with legacy use continuing into the 1980s. I saw Brunsvigas used to calculate the

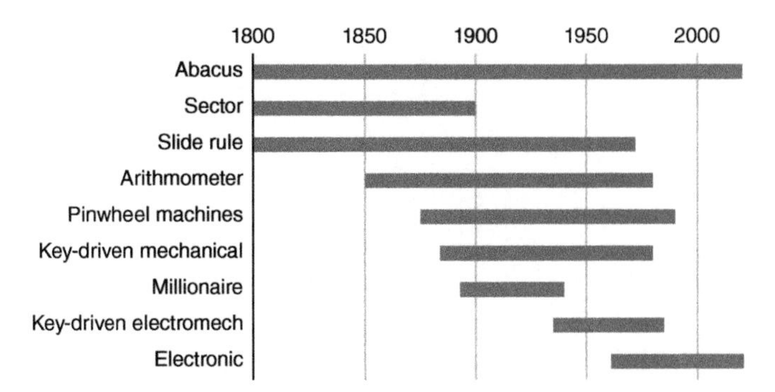

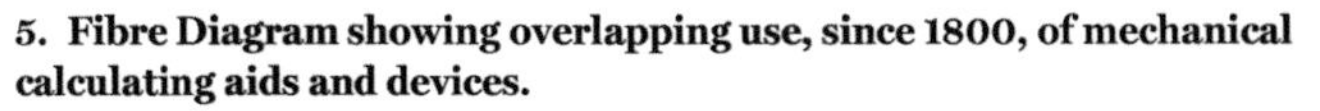
5. Fibre Diagram showing overlapping use, since 1800, of mechanical calculating aids and devices.

odds at the Harringay dog track in North London up to the closure of the stadium in 1987.

The stepped-drum arithmometer, still in use in World War I, was largely superseded by pinwheel and key-driven machines. It made an unlikely comeback in the 1940s in the Curta, an exquisitely engineered cylindrical compact portable hand-held arithmometer. 150,000 Curtas were made by 1972 following their launch in 1948, with their use extending into the 1980s.

The wire-and-bead abacus is still used in parts of Africa, Russia, Eastern Europe, and China. The abacus outlasts all.

Chapter 3
Automatic computation

Automatic computing has its roots in the 19th century. In a single specific event. There is no extended backstory that provides a gentle ramp that prepares us. The lurch into automatic computation was sudden, then faltered.

The genesis episode occurred in 1821, probably in London. Charles Babbage, English mathematician, prominent and controversial man of science, met with his friend, the astronomer John Herschel, to check astronomical tables newly calculated by hand. Dismayed by errors, Babbage exclaimed 'I wish to God these calculations had been executed by steam'. Babbage described this vignette in 1839.

'Steam' can be read as a metaphor for the infallibility of mechanism and also for the automatic production of tables by machinery. The 'unerring certainty of mechanical agency' would ensure tables free from human error. With machine as factory and number as product, mathematical tables, like manufactured goods, could be produced at will—calculated and printed automatically. In Babbage's invocation of steam we have the extension of a model of industrial production from goods to information.

Babbage was 29 at the time, and he spent the best part of the next half-century, until his death in 1871, attempting to mechanize and automate first calculation, then mathematics. His efforts saw the transition from mechanized calculation to fully fledged automatic computing.

His first apparatus was a difference engine, so-called because of the mathematical principle on which it is based—the method of finite differences. A practical advantage of the method is that it allows the values of an important class of mathematical functions, polynomials, to be calculated by repeated cascaded additions without the need for multiplication and division which would ordinarily be required. Addition is much easier to implement using cogs and gearwheels than multiplication or division which were still formidably difficult to mechanize.

His earliest model, made in 1822, was driven by a falling weight. A fully engineered model for demonstration, completed in 1832, was driven by an operator cranking a handle back and forth. Initial values, pre-calculated, were entered by hand. Once set in motion the machine produced, without intervention, a sequence of values, one after the other, of the mathematical function being tabulated. Each cycle of the engine produced the next result, and the user read the sequence of results from a column of figure wheels engraved with the numbers 0 through 9. Designs for a full machine included a printer for inked hardcopy and an apparatus that automatically impressed results in soft material to make moulds from which printing plates could be made, this to avoid the error-prone process of typesetting in loose type by hand.

The operator did not need to understand how the mechanism worked, or the mathematical principle on which it was based, to achieve useful results. The steps of the computational algorithm were no longer directed by human intelligence, but by the internal rules embodied in the mechanism and automatically executed. The algorithm was in the wheelwork.

The experimental model was automatic and represents the first successful embodiment of mathematical rule in mechanism. By expending physical energy one could produce results which up to that point in time could only be produced by mental activity, by thinking. A junior colleague of Babbage wrote that he, Babbage, had 'taught wheelwork *to think*, or at least to do the office of thought'. Autonomous action is the first prerequisite of machine intelligence.

Machine computation

Based on trials with the first experimental machine Babbage speculated on the implications and potential of machine computation.

There are many important 'series' in mathematics—sequences of values defined by a generative rule. If you wished to know the value of the 350th term in a series there was no formal way of finding this unless you had a general expression for the *n*th term. In some instances no such expression was known. By cycling the Engine through 350 values, one after the other, the required value could be found. In this way machine computation provided solutions previously achievable only by analytical methods.

A more far-reaching finding was that the machine could solve equations, especially those for which there was no known analytical solution. For mathematicians the solution of an equation is the value (or values) of x (the independent variable) that reduces the equation to zero. Since cycling the Engine produced each next value of the expression, detecting the all-zero state in the results column signalled a solution. In the case of multiple solutions, continuing to cycle the engine would find each solution in turn. In the event of there being no solutions, the machine continues indefinitely.

Babbage first arranged for a bell to ring automatically when the all-zero condition was detected so that the operator could halt the Engine. He later made provision for the Engine to halt automatically. He wrote explicitly about the machine halting on finding a solution. In doing so he offers automatic computation as systematic method. This was new territory and not much in tune with the prevailing mathematical traditions which privileged analytical methods over 'mere' numerical calculation.

Over a century later Alan Turing, mathematician, code-breaker, and computer pioneer, provided an answer to a seminal question in mathematical logic—whether the conceptual model of mechanical process could determine whether or not a certain class of problems was solvable in principle. The criterion of solvability was whether or not Turing's notional machine would halt. Babbage did not claim any theoretical significance for his finding. But the resonances with Turing's work are unmistakable.

From calculation to computing

Difference engines are not computers as we would now understand the term. They are strictly calculators that crunch numbers in the only way they know how—by repeated addition according to a fixed rule. They execute a single specific algorithm on whatever initial values they are given. In computational terms they have no generality, not even as a four-function arithmetical calculator.

The leap to generality is found in Babbage's Analytical Engine, conceived in 1834. The designs for the Analytical Engine describe a programmable general-purpose machine capable of executing any sequence of arithmetical operations. The Engine features an internal repertoire of functions executed automatically—direct division, multiplication, subtraction, and addition—'micro-programmed' into the mechanisms. The user can program the Engine using punched cards, the hole patterns of which control the machine. The internal organization of the machine pre-echoes signature features of the modern computer described in 1945, some one hundred years later, by John von

Neumann, whose model of computer architecture dominates computer design to the present time.

Babbage's Engines represent a quantum-leap in physical scale and logical complexity in relation to what had gone before. None of his Engines was built in their entirety in his lifetime. Small trial assemblies survive. He failed, notwithstanding independent wealth, generous government funding, decades of design and development, the services of one of the finest tool-makers, and the privileges of a gentleman of science with access to the highest echelons of society and power. He is known as much for his failure to complete an engine as he is for the ingenuity and originality of their invention.

Ada Lovelace

For Babbage the engines were a new technology for mathematics. Until the last, he affirmed that mathematics was the context, and stated object, of his machines. The person who first articulated the wider significance of computing was Augusta Ada Byron, Lord Byron's daughter from his short-lived and troubled marriage to Annabella Milbanke. Ada Byron, later Countess of Lovelace, and Babbage met in 1833, when she was 17 and he 42. Babbage became, and remained, a close family friend until her death in 1852 at the age of 36.

In 1843 Lovelace published an article describing Babbage's Analytical Engine. This paper, 'Sketch of the Analytical Engine', was her only substantial publication. It is on the contents of this single publication that her reputation as a figure in the history of computing rests. 'Sketch' is a translation from French of an article by mathematician and later prime minister of Italy, Luigi Menabrea, who attended a series of sessions by Babbage at a convention in Turin in 1840, the only known occasion on which Babbage spoke publicly about the Analytical Engine. Lovelace appended to the translation her own 'Notes', which are two to three times the length of the translated piece.

The first distinctive feature of the 'Notes' is the shift in the debate from mathematics to the wider significance of automatic computation beyond mathematics. Lovelace speculated that if the rules of harmony and of composition were amenable to computational process 'the engine might compose elaborate and scientific pieces of music of any degree of complexity or extent'.

This statement contains a critical step in signposting the representational power of number. Machines operate on number, that is to say, on physical representations of number—cogwheels, rods, voltages. They cannot operate on notes of music or letters of the alphabet, only on numerical representations of these. A central idea in Lovelace's example is that number can represent entities other than quantity. If we assign meaning to number, then results arrived at by operating on number according to rules can say things about the world when mapped back onto the world using the meanings assigned to them. Lovelace's essential insight was that the potential of computing lay in the power of machines to manipulate, according to rules, representations of the world contained in symbols. Nowhere, in his extensive published and unpublished work, does Babbage write in this way.

Lovelace is also explicit in her framing the Analytical Engine, a general-purpose computer (though she did not refer to it in this way), as different in kind from other machines. 'The Analytical Engine does not occupy common ground with mere "calculating machines". It holds a position wholly its own'. There is recognition here that the automatic general-purpose computer was historically and logically novel. Its generality of use, later formalized by Turing in terms of 'universality', distinguished computing machines as different in kind from 'tools' which have specific and unalterably defined purpose.

Correctives

It is as well to realign received perceptions and popular narratives about Babbage and Lovelace where these are not evidenced. The most frequently cited reason for Babbage's failure to complete any of his engines is limitations of 19th-century mechanical engineering. Assertions of this kind imply, and often directly state, that parts could not be manufactured with sufficient precision for a workable machine.

There is compelling evidence that achievable precision was not a limiting factor. Measurements taken from surviving mechanisms made by Babbage's engineers confirm that repeated parts could be made with a precision of 1.5-thousandths of an inch. Babbage's Difference Engine No. 2, made to the original designs and completed in 2002 at the Science Museum in London, was built with a precision nowhere exceeding what is known from measurement to have been achievable by Babbage himself. Its 8,000 parts, many made less precisely than achievable in the 19th century, function as Babbage intended. The reasons for Babbage's failure are more complex and nuanced than the technological determinism implied by the limits of achievable precision (see Figure 6).

Ada Lovelace is nowadays widely celebrated. Her reputation as 'the first programmer' is one of the most pervasive historical straplines used to describe her role. This attribution is based on the example she used in her 'Notes' for the way in which the Analytical Engine could calculate an important mathematical series, Bernoulli numbers, so-called after the 17th-century Swiss mathematician, Jakob Bernoulli, who was the first to provide a long-sought general formula for the series. Lovelace's Note G contains the 'program' describing the sequence of steps and the internal states of the Engine in the calculation of the Bernoulli series, and it is for this that she is famed.

6. Charles Babbage's Difference Engine No. 2, designed 1847–9. It has 8,000 parts, weighs 5 tons and measures 11 feet long. Modern construction completed in 2002.

If the frequency with which Lovelace is called the 'first programmer' could make it so, her reputation would be unassailable. However, the concept, logic, format, structure, and notation of Lovelace's program are those of the programs written by Babbage, the earliest of which pre-dates the Bernoulli program by some seven years. The last of Babbage's known programs, of which there are some two dozen, was completed in 1840, three years before Lovelace's publication and some two years before she started to work on 'Sketch'. Incontrovertible primary source evidence rebuffs the popular perception that Lovelace was the first programmer, but tabloid history has so far prevailed. It is regrettable that the heavy focus on this depiction has distracted from an appreciation of her more significant contributions.

A final realignment relates to Babbage's influence. Babbage is routinely described as the great ancestral figure in the history of computing. He is often called the father of computing. Genealogical metaphors of this kind overstate what is at best a tenuous connection. There is no unbroken line of influence between Babbage and the modern computer. Most leading figures of the modern era knew of him and the legend of his efforts. A few did not. The principles of modern computing were reinvented in ignorance of the detail of his work, which was not closely studied until the late 1960s. Babbage had no discernible influence on modern computer design. Many of the core ideas of modern computing are evidenced in his work and this congruence raises the question whether these ideas, articulated a century apart, and without direct influence, embody something fundamental about computational process.

The hundred dark years

Not everyone embraced the prospects of machine computation. Many experts were sceptical and were not coy in making this known. When George Biddell Airy pronounced the engines to be 'useless', this was not a single irritable aberration. Airy consistently rejected the practical and economic utility of mechanically assisted calculation, and his views played a defining role in Babbage's fate.

Swiss and French astronomers were similarly jaundiced, variously arguing that existing printed mathematical tables were already sufficiently accurate, and that the high digit-precision of machines was unnecessary given that observational data was anyway inexact. Airy alleged that the demand for computing machines came not from the needs of computers (humans who did the calculations) but from 'mechanists', i.e. from the supply side—machine enthusiasts and those who stood to benefit from their manufacture. In the event, the pragmatic naysayers prevailed and everyone carried on using slide rules, printed tables, and arithmometers.

With Lovelace's death in 1852 and Babbage's death in 1871, the 19th-century movement to mechanize calculation, and generalize automatic computing, lost its two most active protagonists. Babbage's Engines were a false dawn. Automatic digital computing did not revive until the 1940s. Leslie Comrie, who pioneered techniques of mechanical computation, referred to the gap between Babbage and the revival of automatic digital computation in the 1940s as the 'dark age in computing machinery, that lasted 100 years'. The darkness was not total. There were two little-known flickers in the early 20th century, one by Percy Ludgate, an Irish clerk, and one by Torres y Quevedo, a Spanish engineer and inventor. Both produced original designs for automatic calculating machines. Their work is barely known.

Analogue computers

Without machine computation science would stultify from the 'overwhelming incumbrance of numerical detail'. This was Babbage's prediction in 1822. The late 19th and early 20th centuries saw the growth of the industrial movement and with it the mathematical demands of engineering, statistics, warfare, and science. Babbage's prediction began to materialize. A class of computers, analogue computers, relieved some of the pinch-points, and with some success.

Both analogue and digital computing machines are automatic. But they differ fundamentally in the way they represent quantities and how they work. Digital machines represent quantity by a string of discrete digits, and their internal workings operate on numbers (or strictly physical representations of numbers) in a controlled serial process. This has become the dominant model of what we usually mean by computation. Analogue machines represent quantity physically—lowering a weight, the rotation of a shaft, the level or flow of a fluid, an electrical charge. The physical behaviour of the machine is *analogous* to that of the system being modelled. As with slide rules and other scaled devices, analogue machines represent quantities as continuously variable values rather than by discrete (discontinuous) values as in digital devices.

The Phillips economics computer is an example of an analogue computer. The machine, designed in the late 1940s by the economist Bill Phillips, provides a hydraulic analogue for the behaviour of a national economy. Money is represented by coloured liquid. Its levels and flow represent the movement of money—investment, taxation, reserves, public spending, imports, exports—and how these interact. The behaviour of the machine is a physical model of the macro-economic mathematics that describes it. In a sense there is no 'computational process' as understood for the digital model. The machine simply behaves. The 'process' is its uninstructed natural behaviour—its hydraulic grammar dictated by tanks, valves, weirs, vanes, floats, drainage holes, and sluices. And gravity.

Digital machines produce results by calculation, analogue systems, by measurement—the height of liquid in a tank, the time it takes for a tank to empty, the proportion by which income might be diverted (by a vane) between investment and public spending to avoid deficit. The behaviour of the machine tracks dynamic change and gives fixed values at an end-point, if, that is, the economy reaches equilibrium.

Two standout figures in the history of analogue computing are Sir William Thomson, later Lord Kelvin, a Scottish physicist, and Vannevar Bush, an American engineer and inventor. Their work is closely related.

Predicting the tides

The ability to predict tidal rise and fall was a longstanding concern particularly for harbour-masters with responsibility for ensuring sufficient depth to safely dock ships. It was long known that tidal motion was the result of several periodic influences, mainly the gravitational effects of the Moon and Sun, and the rotation of the Earth. There were empirical ways of predicting tides and mechanical forecasting aids. But it was not until the early part of the 19th century that mathematics, through the work of the French mathematician Joseph Fourier, formalized the way in which separate periodic waves could be combined to model almost any waveform.

Kelvin's tide predictor, designed in the 1870s, is an analogue machine that adds together separate tidal components to generate a continuous graph of changing sea levels. The amplitude, frequency, and phase (its relationship in time to other waves) of each constituent motion was found empirically by analysing historical records of tidal depths at a particular location. These harmonic components were represented by a system of gearwheels and pulleys that sum their separate contributions to create the prediction of changing depths. The machine was run by cranking a handle and the results appeared as a continuous curve on a large chart-recorder from which readings could be taken to compile tide tables. Kelvin completed his first predictor in about 1872. Others followed. With their gearwheels, wires, pulleys, and exquisitely engraved dials, one steps into the world of Jules Verne.

Solving equations

There is an important class of mathematical functions, differential equations, that describe the behaviour of a wide range of physical and engineering phenomena—heat flow, electric transmission, ballistics, mechanics, population growth, chemical interactions, economics, astronomy, infectious diseases are some. Solving these equations is invaluable to the point of indispensability to engineers and scientists. Yet some have no formal solution, or have solutions that require prohibitively complex, detailed, or time-consuming calculation. In 1927, Vannevar Bush, in the Department of Electrical Engineering at Massachusetts Institute of Technology (MIT), was 'thoroughly stuck' on some intractable equations of this kind for predicting the behaviour of electric power distribution networks. Bush devised a machine solution, his 'differential analyser', a mechanical system driven by electric motors. His first was completed in 1931.

Bush's analysers were elaborate room-sized table-like devices with long parallel shafts, motors, disc-and-wheel integrators, torque amplifiers, and input and output tables. The Kelvin tide predictor was problem-specific. Bush's solution was more general—his analyser could be reconfigured for different problems, and it provided solutions not readily achievable by other means, if at all. Copies were made for the US, Germany, Norway, and the Soviet Union. Douglas Hartree, English mathematician and physicist at Manchester University, made a trial model in Meccano—a toy construction-set for mechanical model-making. It was unexpectedly accurate with results true to 1 per cent, and a fully engineered version followed. About ten differential analysers were operational before World War II. More were made, for calculating gunnery settings, shell trajectories, ballistics, and predicting the trajectories of German V2 rockets.

The most sophisticated precision mechanical analogue computers were special-purpose and for military use—controlling the aiming of guns, especially for ground- and ship-to-air anti-aircraft fire, and bomb sight computers in aircraft for high-level bombing for which continuous results in real time were imperative.

With developments in electronics, accelerated by war, electronic analogue computers began to rival the evolved sophistication of mechanical systems. The setting up procedure for a Bush-style mechanical analyser was laborious and could take days. In contrast, electronic analogue computers could be reconfigured for a new problem in minutes or hours using patch cables plugged into an array of sockets, as in old telephone exchanges. The ultimate demise of analogue computers came in the 1970s with cost reductions and speed improvements offered by digital microelectronics.

Abandoned past

Analogue computers are largely written out of histories of modern computing. Latter-day histories privilege machines and technologies that can be seen as staging posts in the development of general-purpose digital electronic computing, the central preoccupation of the modern narrative. Insofar as analogue computers are studied it is as a specialist field sometimes included in canonical histories as a gestural courtesy, before hastily moving on to the triumphs of the digital era.

Sidelining analogue machines in this way underrepresents their contemporary usefulness, accuracy, and longevity. Tide predictors using Kelvin's harmonic synthesizers spread to Europe, Argentina, Brazil, the United States, India, and Japan. A United States tide predictor completed in 1911 combined thirty-seven harmonics (compared to Kelvin's original ten) and calculated tides to just over an inch for each minute of the year. It weighed 2,500 pounds, was called Old Brass Brains, and gave service for some fifty-five years, until 1965 when it finally ceded to electronic replacement. Precision analogue computers for military applications continued to be developed into the 1960s and remained in service well after.

Analogue computers were not an idiosyncratic interlude, or the victim of a hasty replacement programme as perhaps implied by their low profile in the historical narrative. They co-existed with, and overlapped digital electronic devices for some forty years, fully spanning the first three generations of digital electronic computers—the wartime and post-war vacuum-tube era till the mid-1950s, the transistor era till 1963, and the first microchip era ending in the early 1970s.

As late as 1948 analogue analysers of the Bush variety were not classified as 'analogue' or as 'computers' but as mathematical instruments, or machines for mathematics. Classifying these machines as 'analogue' acquired currency only when it became necessary to distinguish them from the new digital technologies. The analogue/digital distinction was widely accepted by 1950 and, once established, was projected backwards onto earlier technologies. In this, analogue machines were a casualty of the 'progress narrative' described by the historian Ronald Kline, in which advocates of the new technologies framed 'digital' as the (utopian) future, and 'analogue' as an (inferior) legacy of the past. It was only in the context of the emerging digital electronic technologies that analogue machines were identified as a category that made them visible as a rival technology. And for that they needed a name.

Behemoths

While analogue modelling was one route, the labour, tedium, and sheer difficulty of numerical calculation remained largely unrelieved. Several independent initiatives surfaced in the mid- to late 1930s. Each was a response to the pressing computational needs of physics and engineering, focused, and then magnified, by the imperatives of war. Collectively they signalled a fresh start to automatic digital computation following the less than fruitful efforts of Babbage and others.

The major ignitions were in Germany and the United States. In each case, specific individuals led the charge. These included Konrad Zuse, an eclectically talented civil engineering graduate in Berlin; George Stibitz, a research mathematician at Bell Telephone Laboratories in New York; and Howard Aiken, physicist, mathematician, and engineer at Harvard University. Each had been frustrated by the impediments of numerical calculation: Zuse by the analysis of structures; Stibitz at first by the arithmetic of numbers with two components ('complex numbers'), much used in telephone network design and electrical engineering; and Aiken by an especially challenging set of differential equations for his Harvard physics thesis. In developing problem-specific solutions, each was led to more general-purpose computational machines.

Zuse, Stibitz, and Aiken each started essentially from scratch. Zuse worked from first principles. Self-confessedly he 'neither understood anything about computing machines' nor had any significant knowledge of earlier calculating devices. Based in Germany he was cut off from developments in the US and Britain during the war years, a period of developmental ferment. Stibitz too owed nothing to the traditions or history of mechanical calculation. A study of his work ratifies his own account, 'I have no head for history. I did not know I was picking up where Charles Babbage in England had to quit over a hundred years before'.

Aiken's debt to the past is more problematic. He had read about historic calculating devices and knew of Babbage's Engines, but the depth and extent of his prior knowledge is unclear. His praise of Babbage was extravagant and frequent. 'If Babbage had lived 75 years later, I would have been out of a job,' he told an interviewer. Whatever his knowledge of the past, its influence is difficult to establish with any certainty.

Technology

The period in question is the decade between 1935 and the end of World War II. The instigators of the 'fresh start' pressed into service a variety of technologies: mechanical parts, electromechanical relays, and—towards the end, beckoning from the future—electronic vacuum tubes. For the meanwhile, the new frontier was electromechanical switching relays, of the kind already widely used in telephone exchanges.

The shift from mechanical to electromechanical devices was significant. A relay is essentially an electrically operated switch. Applying an electric current to a coil magnetizes a metal core which attracts a lever that closes or opens a pair of contacts to act as an on/off switch. One simple relay has two states, on or off, and can represent one binary digit (1 or 0).

There were reliability issues. Dust or dirt between two contacts led to failures. In a telephone exchange such malfunction is not overly critical. Redialling would likely route the call through a different path and the next operation of the troublesome relay might anyway clear it. But calculation is less fault-tolerant. A single-bit error, even a transient one, could compromise an entire calculation without the user's knowledge. Relays are bulky. Storing and manipulating large multi-digit numbers requires thousands of them. The computational ambitions of these machines, and the irreducible bulk of the implementation medium, resulted in a brief generation of machines of monstrous size, at least by the standards of the day.

Konrad Zuse's Z-series

Zuse quit his job as a structural engineer at the Henschel Aircraft Company after a few months to become a computer developer, a lifelong dedication. Between 1936 and 1945 he built his Z-series machines, Z1 through Z4, each a staging post for the next. His first two machines, Z1 and Z2, were improvised, built, with help from friends, in the living room of his parents' apartment in Berlin. He wanted a machine that could not only automatically carry out a sequence of calculations but also store and retrieve intermediate results, something most manual calculators were unable to do.

His first prototype machine, the Z1, was entirely mechanical with memory and the arithmetic unit based on sliding metal plates with slots and pins, some 30,000 parts in all. The Z1 sort of worked. For the Z2, Zuse retained the mechanical memory but used relays for arithmetic, this on the suggestion of Helmut Schreyer, a former schoolmate and co-worker. The Z3, completed in 1941, used relays throughout, 2,600 of them, and has the distinction of being the first fully operational, automatically controlled, calculating machine. Schreyer had proposed using vacuum tubes, which were much faster, for an electronic version of the Z3. The idea was abandoned. The military expected that the war would be over too soon to benefit from it.

For the last machine in the series, the Z4, Zuse reprised his well-proven mechanical memory for reasons of cost and size. It was moved several times during the war and concealed underground to avoid capture or bomb damage. When finally installed in Zurich in 1950 it was one of the few operational automatic calculators anywhere.

A feature of all these machines was a degree of programmability by the user. This was a departure from mechanical calculators up to that time. Instructions were automatically read from perforated tape made from discarded 35-millimetre movie film punched by hand. The use of film was again at the suggestion of Schreyer who, while still a student, had worked as a projectionist.

Little was known of Zuse's machines outside Germany during the war and his work had little influence on British or US developments in the post-war years. The use of binary, the internal logical organization of computing machines, and their operational control were invented afresh by others. Zuse went on to a successful career. He held patents, founded his own company, adapted his designs for vacuum tubes and then transistors, and built about 250 computers before his company was finally sold to Siemens in 1967.

Stibitz and Bell Labs

The canonical story of George Stibitz starts at Bell Labs in 1937 when he took home some scrap relays and assembled, on his kitchen table, a small lash-up of a device for adding two one-digit binary numbers. He recalls that his colleagues were 'more amused than impressed' by his Model-K ('K' for 'Kitchen') but the episode led to Bell authorizing work on the Complex Number Computer, a calculator that could add, subtract, divide, and multiply complex numbers (numbers with two components, of the form a + *i*b).

The manipulation of complex numbers is not difficult, but long calculations are tedious. The Complex Number Computer (later renamed the Model I), operational in 1940, was entirely relay-based, using 450 of them. Users typed in calculations on a teletypewriter and the machine printed answers with only momentary delay. The Model I could service more than one user (though not at the same time) and could be used remotely over lines linking the teletypes to the machine. Remote multiple access was a novelty.

Stibitz improved reliability by using additional relays for error detection in much the same way that Babbage used additional security mechanisms to ensure digital integrity a century earlier. Stibitz used a system of binary coding in which each decimal number is represented by seven relays rather than the usual four. This 'biquinary' coding, still called the Stibitz Code, allowed the automatic detection of relay malfunction. Relays are bulky. Storing and manipulating large multi-digit numbers requires thousands of them. Additional relays for error detection were an unwelcome but tolerated overhead.

With the US entry into the war, peacetime research yielded to military priorities, and six digital relay machines of increasing sophistication followed. The main military applications were for aiming and tracking anti-aircraft guns. Though they were special-purpose machines they edged towards generality, this by reading data and instruction sequences supplied on paper tapes that could be changed for different tasks.

The largest and most complex of the Bell Labs machines was Model V of which two were made for military use. Instead of executing only a fixed predetermined sequence of operations it was able to branch, i.e. alter what it next did depending on the result of a calculation just completed. The Model V was not yet a fully fledged computer as we would now understand the term, but it was a productive general-purpose calculator that augured the end of the line in relay machines.

Harvard Mark I

The most ambitious and glamorously extravagant project of this period is the ASCC (see Box 1), later called the Harvard Mark I or simply the Mark I. Howard Aiken, while still an instructor in Harvard's Department of Physics, proposed it in 1937. IBM co-designed, built, and donated it to Harvard in 1944 where it was immediately deployed in the war effort.

The machine was massive—51 feet long, 2 feet deep, it consisted of some 760,000 parts, and weighed 5 tons. IBM paid two-thirds of the $500,000 cost, the rest came from the US Navy. The machine was visually impressive, and deliberately so. Stainless steel and polished glass cladding signalled clinical futurism, a showpiece with corporate publicity in mind. Despite the gloom of war, the popular and scientific press reported the public inauguration. News of the 'giant electric brain' captured attention and boosted public awareness of computing machines.

The thrust of Aiken's case, articulated in his written proposal, was that scientific calculation was not well served by office accounting machinery of the kind already in widespread use. He argued that scientific calculation had a greater role for negative numbers (presumably debt was not presumed to be normative) and for a range of mathematical functions. Perhaps most importantly, scientific calculation required iterative processes, that is, the automatic repetition of the same calculation on different data, for tabulation, or for successive approximation, for example. Aiken's proposal did not call for revolutionary technology but for using a versatile arrangement in which numerical data could be routed through different pieces of already available apparatus.

The various arithmetic units were based on standard IBM accounting machine equipment, a market IBM dominated. The whole was a mix of electrically driven mechanical and electromechanical components with over 5,300 miles of wiring. It clicked and clacked when it ran, 'like a roomful of ladies knitting'.

The Mark I's wartime work included calculations for magnetic fields associated with magnetic mines for ships, radar technology, and computations for the design of the atomic bomb developed at Los Alamos, New Mexico. After the war it was used for a variety of research projects including language translation and economic modelling, perhaps the first application of a computer to the social sciences. It was also used for the more mundane task of calculating and printing mathematical tables. Automatic typewriters printed out the columns of results which were reproduced using a photolithographic process.

Here, finally, was a machine that produced printed mathematical tables and eliminated human fallibility from the chain of processes ordinarily involved in tables production by hand—calculation, transcription, typesetting, printing, and verification. The typography was plain, without the elegance or ornamentation of 18th- and 19th-century manual typesetting using loose type, but the final product had what Babbage had envisaged, 'the unerring certainty of mechanical agency'. 'Babbage's Dream Come True' was the title of a piece by Leslie Comrie published in *Nature* in 1946. It had taken a full century.

Bugs

There is not much humour let alone whimsy in computer history. So the story of the computer bug must be told. Aiken supervised the design of the Mark II, the first of three machines to follow the Mark I. One day in 1947, the year it was completed, a fault arose and the malfunction was traced to a moth that had got stuck between a pair of relay contacts. The hapless moth was taped into the logbook above an entry 'First actual case of a bug being found'. 'Bug' had been in use at least since 1878 by Thomas Edison, as a name for a technical fault or glitch. But this well-evidenced folkloric tale, based on a pun, has been widely adopted as the origin story of the use of 'bug' and 'debugging' for faults, fault-finding and fixing malfunctions in programs, hardware, and engineering.

Grace Hopper, a newly commissioned naval officer, assigned to Aiken's team in 1943, made the logbook entry. Hopper had a lifelong career in computing, making major contributions to computer programming and computer software. She wrote the first widely used compiler—a program that translates instructions, written in abbreviated English (source code), into the internal codes for execution by machine. She was also instrumental in the development of the computer language COBOL, widely used for business applications.

Positioning the Mark I

The significance of the Harvard Mark I is both symbolic and practical. As a versatile, completed, visually imposing, fully automatic computing machine, it came to symbolize the prospects of the new computer age, both in public perception and in scientific communities. In technological terms it was a dead end soon to be superseded by electronics. As for Aiken himself, his influence transcended his machines. The Harvard Computation Laboratory—with Aiken, a committed teacher, as director—was the training ground for a generation of engineers, designers, and practitioners that went on to influence business, military, and academic computing in the decades that followed.

The machines of this period signified a fundamental shift from analogue computation, with perpetual anxieties about precision, to digital numerical computation where imprecision is exactly quantified, i.e. a number is precise only to the value of its last digit and no more, and is agnostic about the value of unrepresented digits beyond it. The shift is one from measuring to counting, from analogue continuity to the discretized world of digital switching, a central feature in what was to come.

In the narrative of the Harvard Mark I, historical agency is shared between the 'heroic inventor' (Howard Aiken) and a corporate entity (IBM). The names of those prominent in the IBM team are known and documented (James W. Bryce, Benjamin M. Durfee, Francis E. Hamilton, and Clair D. Lake) but it is IBM and its chairman, Thomas J. Watson, that are foregrounded alongside Aiken. As the computer graduated from the lab to the market, corporations feature more prominently in accounts of historical agency, with individuals increasingly anonymized. There is an irony here in that Watson, who headed the towering corporate entity that was IBM, celebrated 'the power of one'—individual inventors who single-handedly innovated products for manufacture and sale by corporations.

Chapter 4
Electronic computing

The appeal of vacuum tubes was speed. Vacuum tubes, as they are called in the US (known as valves or thermionic valves in the UK) are small glass vials, evacuated of gas, containing metal plates (electrodes). Electrical signals, applied to the electrodes, control the flow of current through the device.

Before they were used in calculators and computers, vacuum tubes were used in radar and radio equipment as well as in audio circuits to amplify continuously varying signals as, say, from a microphone. Calculators and computers, on the other hand, used vacuum tubes as digital switches which are in only one of two states at any time, on or off, with negligible time taken to flip between the two. New techniques and circuits were needed to use vacuum tubes in this unaccustomed role. Electronic switching is fast. Vacuum tube calculators in the 1940s were 500 to 1,000 times faster than designs using electromechanical relays. Vacuum tubes have no moving parts. There is no mechanical wear.

Speed was the prize. But computational speed was only one element of design. To take advantage of electronic speeds required fast retrieval of information from memory. Storing information for fast retrieval was a stubborn challenge and some of the techniques were as unlikely as they were exotic. Getting information into the machine, and results out of it, needed to be fast enough not to squander the hard-won speed advantages of electronic calculation. Vacuum tubes had a reputation for unreliability, and this was a major early deterrent to using large numbers of them. How machines might be logically organized (their architecture) was also far from settled. And the difficulty of altering what the machine did (programmability) soon became a defining impediment.

There were several independent initiatives in Britain, Germany, and the United States. The three machines most prominent in the narrative are the ABC, Colossus, and ENIAC (see Box 1). Each addressed design issues in different ways.

The ABC computer

John Vincent Atanasoff, a physics professor at Iowa State College, with a first degree in electrical engineering, was not the first to be daunted by manual solutions for solving equations. In 1937 Atanasoff conceived the essential features of a machine solution. It would be electronic (using vacuum tubes), digital, and use binary for its internal logic.

With the help of Clifford Berry, a graduate assistant, Atanasoff completed a prototype in 1939 and a full-scale machine in 1942. The machine was not programmable nor fully automatic. The operator needed to intervene to guide the process of solution to each next stage by manipulating switches, push buttons, and cards. But it was a successful design of a special-purpose fixed-program digital electronic calculator for the solution of certain equations, notwithstanding some stubborn reliability issues.

Both Atanasoff and Berry left Iowa in 1942, and their desk-sized calculator was effectively abandoned and forgotten. Were it not for a patent dispute and a high-profile court case that came to judgement thirty years later, the ABC, a departmental university project with a modest budget of $5,000, would surely not be part of the core narrative. At best it would be included as a homage to precedent, a toe dipped in the water, with little practical use, and no material consequence to the way the story came to be told. As it happened the ABC is an epilogue to the story of ENIAC, one of the trilogy of canonical machines.

Code-breaking and Colossus

Colossus was a class of code-breaking machines built between 1943 and 1945 for use at the Government Code and Cypher School at Bletchley Park, a large country estate some 50 miles outside London. Colossi were built to help decipher intercepted German radio messages encrypted by the Lorenz SZ40 cipher machine used for top-level communications between the German high command and army group commanders. Decrypted cipher traffic yielded invaluable military intelligence, including German troop positions critical to the Normandy D-Day landings and confirmation that the Allied deception and deliberate misdirection of the intended location of the D-Day landings had succeeded. The Colossi were designed and built at the Post Office Research Station, Dollis Hill.

The Lorenz machine is an attachment connected to a teleprinter. It automatically encrypted the stream of pulses produced as the operator typed the outgoing message. The encrypted text was transmitted, usually by radio, to a recipient equipped with a second Lorenz machine set up by the operator with exactly the same code settings as the sender's machine. The Lorenz at the receiving end decrypted the incoming pulse stream to recover the original plain-text message. The Allies intercepted the transmitted signals in their encrypted form. Remarkably, without sight of or access to the machines themselves, Allied code-breakers worked out the logical structure of the German machine, recovered the keys used for the encryption, and recreated the original plain-text message.

Cracking the Lorenz traffic was a formidable feat of intellect and technology, involving mathematicians, cryptanalysts, and technicians at Bletchley Park working in collaboration with engineering staff at Dollis Hill. A cardinal error by a German cipher operator (resending a long message without changing the settings) opened a chink in Lorenz's otherwise near-impenetrable defences.

Breaking the Lorenz settings was not a once-only task. German operators continually changed the settings, and the codes needed to be broken afresh each time. The number of combinations of the encryption key is astronomically high and the time to try all possible keys one at a time would be a useful way of defining eternity. Wartime intelligence becomes operationally irrelevant after a very short time. It was a question of minutes and hours rather than days or weeks.

Heath Robinson

Colossus's predecessor was called Heath Robinson, after the elaborately improbable cartoon contraptions depicted by the English illustrator of the same name. The Robinsons had two loops of paper tape punched with holes. Tapes were up to 200 feet long, looped back and forth around a series of pulleys in a frame called the Bedstead. One tape held the enciphered text, the other, a part of a candidate key to be trialled. Each trial required a complete pass of the two tapes, and for each run the tapes were automatically shifted one position in relation to each other. The tapes were read by optical sensors as they ran through the readers at around 30 miles an hour. The optical readers produced a stream of electrical pulses from the hole patterns on the tapes and the machine automatically evaluated the streams to find the most promising match.

The Robinsons generated the stream for the trial key by reading the key tape. Colossus generated this stream electronically. So only one tape, the one with the cipher text, was now needed. This removed the difficulty of keeping the two tapes exactly synchronized (this had been problematic) and saved having to prepare a new key tape for each trial as the settings were now directly configurable using switches and plug boards. There was a further advantage: not having to keep the two tapes aligned allowed the single remaining tape to be run much faster. The benefits of Colossus were a mix of speed, reliability, and operational efficiency.

Colossus electronically correlated the relationship between the two streams, one from the tape and one generated electronically, and produced a count as a measure of likelihood of the correctness of the part of the key being tested. These results were then taken further by others in the next stage of the decryption process, much of which was manual, laborious, and exacting.

The original Colossus was built under the direction of Tommy Flowers, a telecommunications engineer at Dollis Hill, with extensive experience of using vacuum tubes in telephone networks. This first Colossus, built in under a year, was working by the end of 1943 and an improved version by mid-1944. By the end of the war ten Colossi were running with one further in progress. The later Colossi had 2,400 vacuum tubes and 800 relays, and each filled a largish room. The original drawings were destroyed and no wartime Colossi are thought to have survived. A working reconstruction of Colossus was completed in 2007 at The National Museum of Computing at Bletchley Park.

No electronic machine with that many valves had been attempted up until that time, and the success of Colossus established the viability, in a computational context, of large-scale use of vacuum tubes, about which many had been sceptical. It is questionable how far this knowledge spread beyond the bubble of Bletchley secrecy. The influence of Colossus migrated into peacetime computing through the expertise of those who designed and worked on it, many of whom became leading figures in post-war computing.

Fiction and fact

Many champions of Colossus describe it as the 'the world's first electronic computer'. A Royal Mail commemorative stamp issued in 2015 has the text 'COLOSSUS world's first electronic digital computer'. These claims have always made me uncomfortable.

The public narrative follows the favoured tropes of tabloid history, many of which underpin the movie *The Imitation Game*, in which the misunderstood genius is pitted against impossible odds, invents the electronic computer, and wins the war. The movie, released in 2014, is a gratuitously fictional account based (sort of) on the efforts of Alan Turing (played by Benedict Cumberbatch), mathematician and Bletchley wartime code-breaker, in an inglorious mashup of myth, occasional fact, wishful thinking, and improbable tosh.

Secrecy about the Bletchley Park decryption operation extended well after the War. The information vacuum was filled by anecdote, speculation, and fragmentary witness accounts that were difficult to corroborate given that there was so little technical detail in the public domain. Inevitably, there were contradictions and distortions. It took decades to loosen wartime restrictions on disclosure. The existence of Colossus was confirmed in 1975 after thirty-two years of official silence, with the release of a set of captioned photographs. Colossus's protagonists sought to secure recognition for the machine. They chose the unquestioned currency of merit, distinction, and historical significance—the extent to which a machine can be framed as a prototypical programmable electronic computer. Colossus has been paraded this way ever since.

I have always felt that presenting Colossus in this way does Colossus and history a disservice. It puts the machine in the position of having to impersonate a general-purpose computer to justify a place in the hall of fame, and creates the awkward predicament of a guest at a celebrity party, anxious that they might be there under false pretences.

A new study historicizes Colossus more fully. Its authors, computer historians Thomas Haigh and Mark Priestley, consider how history has so far framed the machine, and whether Colossus, a special-purpose machine for decryption of Lorenz traffic, that did logic and bit-stream processing, can meaningfully be categorized as a computer, and in what sense it might be thought to be programmable, as has been claimed.

For historians, Colossus is an outlier. It does not fit comfortably in the thread of automatic computation taken as numerical calculation, the role that came to define what was meant by a computer at the time. The overall context of Colossus is communications—German high command communicating, using encrypted radio messages, with officers in the field. High-speed tape reading, trial-and-error comparisons, automatically generated trial keys, all fit well in the information management thread. As well as automatic decryption techniques using Colossus, Bletchley code-breakers used manual methods—graphical, tabular, logical—devised to reveal hidden informational structures and patterns. These too have a natural home in information management. So in the four-thread scheme described earlier, Colossus is better placed in the communication thread and in information management than in automatic computing. Colossus is properly a guest at the party, but in its own right.

Problems with guns

The problem was artillery firing tables—printed booklets or pamphlets for use on the battlefield from which guns could be set correctly for shells to strike their intended targets. The trajectory of a shell depends on many factors—the elevation of the gun barrel, shell shape, shell weight, explosive charge, distance to the target (range), wind strength and direction, angular velocity of the Earth, air density and temperature, both of which are altitude dependent. Anti-aircraft shells had timed fuses that had to be pre-set to detonate at a particular point in the predicted trajectory close to the target aircraft. Performing calculations in the field to take such factors into account was out of the question. Printed booklets contained pre-calculated look-up tables for gunners to make the final settings.

Tables were calculated by solving the trajectory equations for each specific pairing of shell and gun. For even the most basic settings a usable firing table required from 2,000 to 4,000 trajectory calculations. Human computers using desktop calculators took several months to produce one table for a full range of settings and corrections. Electromechanical differential analysers took about thirty days for a table, and a separate table was required for every combination of gun, shell, and fuse. The equations were known and solvable. It was a question of computational speed and throughput.

ENIAC

War was a catalyst. New applications, critically urgent timescales, unforgiving reliability, development unshackled from the expectation of commercial return, and priority access to material and human resources were all accelerators.

The US entered the war in December 1941. The Ballistic Research Laboratory (BRL) of the US Army recruited a hundred, and then a further hundred, human computers, mainly women, and provided training to carry out the trajectory calculations using desktop mechanical calculators. As the war intensified new types of artillery escalated demand for firing tables. By the start of 1943 the BRL was fighting a losing battle. By mid-1943, the situation was critical.

Calculating at electronic speeds was a compelling prospect. In June 1943 the BRL signed a contract with the Moore School of Electrical Engineering at the University of Pennsylvania for the construction of an automatic electronic calculator for ballistics calculations. Three of the prime movers in the proposal for an electronic solution were John Mauchly, a physicist with electronics training and a professor at the Moore School; John Presper Eckert, Moore School instructor and electronics engineer; and Lieutenant Herman Goldstine, a mathematics PhD and ordnance mathematician at BRL. While the Army had firing tables as its specific goal, Mauchly and Eckert, and others at BRL, had ambitions for more general use than ballistics.

The contract was for the Moore School to build the ENIAC for the Army. Mauchly was the 'visionary', Eckert the committed and meticulous driver of hardware design, and Goldstine the main liaison between the US Army and the Moore School. In physical scale and logical complexity ENIAC was ambitious beyond anything yet conceived. Its circuitry contained over 18,000 vacuum tubes, 1,500 relays, 6,000 switches, 70,000 resistors, and 10,000 capacitors. It weighed 30 tons and had two 20-horsepower blowers for cooling.

Reliability

Not everyone was enthusiastic. The main reason for scepticism was the unreliability of vacuum tubes. Eighteen thousand vacuum tubes were many times more than anything anyone had yet tried, and building a reliable machine from unreliable components seemed an engineering contradiction.

Eckert and his team of about a dozen young engineers developed techniques and strategies to reduce failures. Vacuum tubes have heaters—filaments, as in a traditional light bulb—that release electrons when hot. It was known that vacuum tube heaters are most vulnerable to failure when switched on from cold because of the large, though brief, current surge. So ENIAC was never switched off. Running a vacuum tube below its standard ratings extended its operational life so they were routinely underrun. Circuit components including resistors, and especially vacuum tubes, were pre-tested and pre-selected to ensure that only those well within their rated manufacturing tolerances were used. Circuits were on plug-in units for rapid and easy replacement if they failed. The suite of measures was largely effective. Uptimes were routinely over 90 per cent. Goldstine reported runs of three days without error.

Architecture

ENIAC played a defining role in the realization of what was to become the generic electronic general-purpose digital computer. But it differed significantly from the near-universally accepted model that was to emerge. Its conceptual roots lay in mechanical calculators and in Bush-type differential analysers. Arithmetical processing was not centralized in a single functional unit but distributed in separate dedicated arithmetic units. ENIAC used decimal rather than binary, though the way ENIAC implemented decimal used more vacuum tubes than minimally necessary. Redundancy increased reliability. The main reason for preferring decimal to binary was to avoid double conversion into and out of binary. It was an expensive choice.

Setting up ENIAC to do a particular calculation was no picnic. The process involved cabling its separate units in a specific configuration. This was done by plugging hundreds of cables into plug boards and setting banks of switches. Firing-table calculations repeat the same calculations time and time again on different data, and once set up, changing input data was relatively fast. But to solve a new problem meant replugging the whole machine. It took about two days to set up ENIAC for a new problem, and planning a new configuration could take months. The machine had a degree of generality, but exploiting it involved laborious reconnection.

ENIAC was big, expensive, and fast. A differential analyser took fifteen to thirty minutes to compute one trajectory. ENIAC took twenty seconds. But too late for the war. Working double shifts the Moore School team completed ENIAC late in November 1945. Germany had surrendered in May.

The Moore School formally inaugurated ENIAC on 15 February 1946. Hundreds of flashing panel lights signalled arcane and wondrous internal processes. The media hailed ENIAC, capable of thousands of operations a second, as a giant 'electronic brain', a phrase tiresomely beloved of journalists. Unveiling the machine, secret till then, sparked scientific and public interest. ENIAC, with its unprecedented speed, size, and complexity, became identified with the start of the electronic computer age.

Situating ENIAC

ENIAC has been heralded as 'the first programmable general purpose electronic digital computer', though, until recently, with ambivalent conviction. It has also been more modestly described as a special-purpose electronic calculator framed as a transitional machine, a staging post, to the modern fully fledged general-purpose electronic digital computer, seemingly the only terms in which it is permissible to confer honours. Situating it remains conflicted.

The ENIAC narrative habitually fades soon after the machine was completed, in December 1945, i.e. before it did much useful work. Yet in the early post-war years ENIAC was the only fully working electronic computer in the US and its services were in immediate demand. No longer constrained by the narrow needs of calculating firing tables, its capabilities and operational practices were extended, enhanced, adapted, and elaborated in response to the needs of new problems and its own discovered shortcomings. It was quickly used in the design of the hydrogen bomb, part of the Manhattan Project at Los Alamos. ENIAC was a valued workhorse in the immediate post-war period. But by the early 1950s it was competing with a new generation of machines. ENIAC was retired from service in October 1955 with no ceremony or fanfare.

A recent study more fully excavates ENIAC's post-war afterlife and largely resolves conflicting claims about its defining capabilities. The new analysis, by Haigh, Priestley, and Crispin Hope, seeks to demonstrate that, as it evolved, ENIAC had the essential functionality, incipient or mature, that defines the modern machine. What has impeded the historical framing of ENIAC is the way its capabilities were implemented. By later standards the machine was too idiosyncratic to be identified as belonging to the modern computer paradigm, which became the terms in which historical influence came to be judged. Their findings suggest that the claims made for ENIAC by its protagonists are not, after all, an overreach, and are largely vindicated.

As guest curator at the Computer History Museum in California I worked on the conceptual treatment for an exhibition then titled 'Computing: The First 2000 Years'. We drew up a list of ten seminal machines in no particular order and then, one at a time, removed the machine that ranked lowest by perceived historical importance. What remained was a list of one. ENIAC.

A chance meeting

Goldstine was waiting for a train from Aberdeen, Maryland, the location of the BRL. It was summer 1944. There he bumped into John von Neumann, Hungarian-born mathematician and physicist, whose contributions to these fields had already earned him a formidable reputation as one of the great minds of his generation. This chance meeting had profound consequences.

Von Neumann was involved in several war projects, one of which was the atomic bomb programme. He had exhausted available methods of computation, including human computers using mechanical calculators. He had visited Aiken at Harvard and done trial calculations on the Harvard Mark I. He had visited Stibitz at Bell Labs, and had IBM punched-card accounting equipment upgraded and pressed into service for scientific computation. Von Neumann was exercised by the need for computational capabilities to model detonating an atomic bomb by imploding fissile material. Goldstine's description of ENIAC on the Aberdeen railway platform, and the prospect of a racy 333 multiplications per second, hit the mark.

Von Neumann joined the Moore School project as a consultant in autumn 1944. Eckert and Mauchly briefed him on the ENIAC design, frozen a few months before, and also on their ideas for a successor machine, the more general EDVAC (see Box 1). Von Neumann met with Eckert and Mauchly every few weeks until early 1946 and participated in the deliberations about EDVAC's design. The team culture was open and co-operative. Ideas and expertise were shared freely.

In a document dated 30 June 1945 von Neumann described the logical structure and operation of a new class of machine. 'A First Draft of a Report of the EDVAC' is a 101-page synthesis of ideas and techniques incubated in the ENIAC project, and is arguably the most influential single document in computer history. The idealized model it describes became the seminal template for computer design and largely remains so.

The report was unclassified. It was initially intended for internal circulation, and Goldstine circulated thirty-one copies for comment. Dissemination was uncontainable and the report, though a draft with missing references and blank spaces, was eagerly sought.

Contested provenance

Though the report grew from the shared ideas of the ENIAC/ EDVAC team, captured and formalized by von Neumann, it bore only von Neumann's name. The 'von Neumann machine' or a machine with 'von Neumann architecture' became the generic descriptor of the modern computer. Whether intentionally or not, by citing only his own name as author, von Neumann appeared to be diminishing or even ignoring his co-inventors, undervaluing engineering design and implementation, and claiming sole credit. Resentments exacerbated pre-existing tensions between Eckert and Mauchly, the pragmatic engineers on one hand, and von Neumann and Goldstine on the other.

It was not only a question of academic credit. Provable provenance affected patent prospects, and Eckert and Mauchly had commercial ambitions not shared by von Neumann. The Moore School group broke up in March 1946 when Eckert and Mauchly resigned after refusing to accept new requirements to assign future patent rights to the university. The pair left to found their own computer company, the first to offer electronic computers as commercial products. Von Neumann went on to establish the Electronic Computer Project at the Institute for Advanced Study (IAS), Princeton. A clutch of computers was the outcome of the IAS work, each with acronymic names of variable intelligibility. JOHNNIAC, in homage to Von Neumann, and MANIAC, are two (see Box 1).

The Moore School patent standoff that led to Eckert and Mauchly's walkout was not the only dispute about rights and precedence. The inventive source of ENIAC was challenged in court. In June 1941 John Mauchly paid John Atanasoff a five-day visit during which he was a guest in Atanasoff's home in Iowa. Atanasoff and his assistant, Clifford Berry, shared details of their machine and its design. The question of whether Mauchly was or was not influenced by his visit to Atanasoff was central to a complex patent dispute. The trial involved 135 days of courtroom sittings during which seventy-seven witnesses gave oral testimony, including Mauchly and Atanasoff.

The verdict, delivered in October 1973, voided Eckert and Mauchly's patents. The judge found unequivocally that basic ideas of ENIAC's design were derived from Atanasoff and the Atanasoff-Berry Computer. His ruling was decisive: 'Eckert and Mauchly did not themselves first invent the automatic electronic digital computer, but instead derived that subject matter from one Dr John Vincent Atanasoff'.

The case was an infamous episode of disputed influence that split the community. There were those who acknowledged Atanasoff's priority. There were others who, despite the court verdict, favoured the work of Eckert and Mauchly as the defining contribution to the realization of ENIAC. To those indifferent to issues of intellectual property rights, academic credit, or the substantial commercial interests (licence fees and royalties), the rumbling dispute ranked as an unseemly scramble for the plaudits of history.

The court ruled on events that had taken place over thirty years earlier and the outcome dented the narrative of Eckert and Mauchly's priority that had prevailed in the meantime. The court had categorically transferred priority from Eckert and Mauchly to Atanasoff. Gordon Bell, American computer scientist and major figure in minicomputer design, commented that the ruling had 'disinvented' the computer. For historians, the complex lawsuit did little to disaggregate individual contributions, which remain debated. Yet it is Eckert and Mauchly who remain pre-eminently identified as ENIAC's progenitors.

Internal stored program

The feature of the new class of machine that history has singled out as its signature is the 'internal stored program'—holding the program in fast-access internal memory from which instructions can be retrieved and executed at electronic speed.

The stored program feature grew out of one of ENIAC's main shortcomings—the difficulty of changing its operation each time there was a new calculation to perform. Instead of having to physically reconfigure the machine and tolerate the downtime involved in setting up afresh, in the new model the sequence of instructions (the program) would be prepared separately off-line on an external medium—paper tape or punched cards—and read into electronic memory. When the program was run, instructions would be executed from this internal source at electronic speeds. The electrical configuration of the machine would remain fixed. The machine would be 'programmable'—its function could be altered simply by changing the program, i.e. loading up a different set of instructions—without any change to physical configuration. Being able to read and execute a program in this way is unremarkable to us. At the time, it was ground-breaking.

Popular and scholarly accounts routinely cite the stored program as the defining feature of the modern computer. The prevalence of this statement, made with little if any explanation, suggests that the significance of the stored program must be self-evident. After many years of puzzlement I began asking around. I asked leading computer historians, first-generation programmers, early computer practitioners, Stanford University computer science PhDs, and Silicon Valley entrepreneurs—what was it about the stored program that allows it to occupy its central defining role? There was no single answer. Speed and operational convenience was one immediate response. There were others. The ability to store instructions and data in the same memory space allowed memory resources to be allocated favouring one or the other depending on the needs of the problem. Programs could modify themselves to create jumps and branches. Self-modifying programs were to have significance for self-learning programs, and AI.

With programs and data in the same memory space, one program could operate on another program treated as data. This had far-reaching implications for compilers—programs that bridge the usability gap between human programmers, who are at home with textual language, and the machine's internal (binary) language that is native to it, but unintuitive to us.

This confusion about the significance of the internal stored program feature is not clouding an earlier clarity. There is no single identifiable inventor whose writings might resolve intended benefits or provide a comprehensive rationale. Issues of priority and precedence are still debated. The 'First Draft' nowhere uses the term 'stored program', nor does it offer a catalogue of operational benefits or theoretical potential. Other descriptors were used for the new machine—an 'EDVAC-type' machine, or a machine with 'von Neumann architecture'. The implications of the internal stored program were far from fully foreseen at the time. 'Stored program' become a shorthand for a mix of techniques, benefits, and implementation technologies that evolved with time.

While each of the answers from those I questioned illuminated a different trait in a tangle of principle and practice, there was one feature about which all were agreed. No one challenged the status of the internal stored program as the defining feature of the modern electronic computer.

What happened next

Two strands dominate the next phase of the narrative up to the mid-1950s. The first is how word spread—a diffusion narrative about how the new computer paradigm propagated and seeded major development projects in the post-war period. The second strand is technology—the development of memory, programming practices, and principles.

Word was first propagated by a procession of visitors to the Moore School eager for information. Then came the spreader event—a course of summer-school lectures.

Press and newsreel coverage of ENIAC's unveiling, and technical articles in engineering and scientific journals, stoked worldwide interest. Requests to the Moore School for information, guidance, and work-placement collaborations rolled in from government agencies, laboratories, industrial research centres, and university departments. Leading figures from Britain visited, a non-trivial expedition in the immediate aftermath of the war. Prominent mathematicians, and luminaries of machine computation included Max Newman, Leslie Comrie, Douglas Hartree, Maurice Wilkes, and John Womersley. All played significant roles in post-war computing. Insider information, much of it still restricted, was prized. Comrie, one-time superintendent of the Nautical Almanac Office, who visited the Moore School in spring 1946, had a coveted copy of the 'First Draft'. The eight-week summer school in July and August 1946 lifted the lid off residual secrecy. Key figures in the EDVAC team—Eckert, Mauchly, von Neumann, and Goldstine—returned to lecture. Howard Aiken joined them from Harvard.

The first generation

The 'First Draft' described a model of a new class of machine. But it did not contain details of how it was to be physically realized. ENIAC had only twenty words of memory implemented using vacuum tubes. Scaling up would take a prohibitively large number of tubes and there was as yet no proven alternative.

At least three projects attacked the problem, all seeded by the Moore School work, two in the UK at Manchester and Cambridge, the third at Princeton, New Jersey. By the standards of modern solid-state devices, the technologies explored seem outlandish. At Manchester University, Freddie Williams, formerly a radar engineer at the Telecommunications Research Establishment (TRE), developed what came to be called Williams-tube memory. The device was based on the observation that the phosphor coating on the inside of a cathode ray tube screen of the kind used for radar retained its charge for a short while before decaying. So if an electron beam bombarded a spot on the screen, the glow of the dot persisted after the bombardment had stopped. Without intervention the spot would fade to nothing. But if constantly read, and its presence refreshed, the charge could be held indefinitely.

The 'Baby'

The Williams tube stored information as a pattern of charged dots on an area of the screen. Each dot position could be written to and read from, and the pattern was constantly refreshed. Williams, with his assistant, built a small no-frills experimental prototype computer to test the new memory by executing a live program. The design of the SSEM (see Box 1) was guided by Max Newman, professor of pure mathematics at Manchester, who had played a key role in the Allied code-breaking efforts at Bletchley Park and had himself visited the Moore School in 1946. The Manchester 'Baby', as the machine came to be known, executed the first electronic internally stored program on 21 June 1948.

The Cambridge EDSAC

When it came to fast-access storage, Maurice Wilkes, director of the Mathematical Laboratory at the University of Cambridge, took a different tack. Wilkes had attended the tail end of the Moore School lectures as an invitee. Before leaving for the US he had seen a copy of the 'First Draft', lent to him by Comrie who had himself just returned from a visit to the Moore School. Wilkes' brief as head of the Mathematical Laboratory was to set up and provide central computing facilities and services for university research. He returned to Cambridge from the Moore School summer school in September 1946 'fully genned up' and convinced that the stored program was the future. He set about designing and building a practically usable machine and was the driving force and principal designer of the Cambridge project.

The 'First Draft' uses the concept of delay memory. This was based on work by Eckert and Mauchly, who planned to slow down the transmission of pulsed signals by sending them down a tube of mercury. Sound travels more slowly than electricity so electrical pulses were converted to sound and propelled down the tube. The time to travel down the tube of mercury provided a short delay. At the receiving end, the train of acoustic pulses was converted back into electrical pulses, amplified, restored to their ideal shape, then rerouted back to the start. The pulse string could circulate indefinitely, trapped, as it were, in this closed circuit and be available to be read and rewritten. Wilkes pursued this route with success. Thirty-two tubes of mercury (called 'tanks'), each 5 feet long, could collectively store 1,024 18-bit words, not much by modern standards but a bounteous improvement on ENIAC.

The use of delay-line storage is enshrined in the name of the machine—EDSAC (Electronic Delay Storage Automatic Calculator)—an overt homage to EDVAC. On 6 May 1949 EDSAC read a program from punched paper tape into memory, computed, and printed on a teleprinter a table of squares of the first hundred integers exactly as instructed. So the first usable EDVAC-type computer was not EDVAC but the Cambridge EDSAC—a difference of one letter, a continent, and a podium place as a world first.

EDSAC demonstrated the viability of the stored program computer. But its historical significance extends well beyond a practical proof of principle. EDSAC laid the foundations of programming practice. Wilkes's explicit purpose was to support the computational needs of university research which the Mathematical Laboratory had till then met with a differential analyser, punched-card calculating equipment, and desktop mechanical calculators. Technical viability was only the first step. The machine had to be reliable, which it sort of was, but moreover usable by other than its highly specialized makers. The Laboratory would provide advice and instruction but users would be expected to do their own programming. This was entirely new territory.

Software

Early on Wilkes was committed to the principle that programs should be written in a notation meaningful to humans. A letter of the alphabet signified function—A for Add, **S** for Subtract, **T** for Transfer, for example, though most did not gesture the function they stood for (**L** for Shift, **V** for Multiply…). Also, that the conversion of the programmer's symbolic instructions into the internal binary language of the computer should be done automatically by the machine so that the programmer is spared having to code in binary, which is less intuitive.

Wilkes had assembled an exceptionally talented team, prominent among whom were David Wheeler and Stanley Gill, taken on as research students near the start of the project. Wilkes, Wheeler, and Gill pioneered programming principles and systematized programming practice. One of Wheeler's many defining contributions was the use of subroutines—free-standing programmes that did specific tasks and could be called on by the 'master' program. Subroutines tended to be generic utilities, such as printing a result, reading an input tape, performing a variety of mathematical processes, or program checking. The subroutine library became a cumulative repository of tried and tested routines available to programmers, and subroutines remain a basic feature of most programming systems. A second major feature was diagnostics—techniques and routines for verifying program code and checking correctness. Programs need to be uncompromisingly exact to work properly. They rarely worked first time and detecting errors could be baffling, frustrating, and time-consuming.

The EDSAC team openly disseminated knowhow and practices through supervision, hosting research placements, publications, fortnightly colloquia, and annual summer schools for all-comers. Major figures in a variety of academic disciplines, laboratories, industry, researchers, and students had access to a leading centre of expertise for computing technology, and for computational practices for research. Wilkes, Wheeler, and Gill published the first textbook on programming in 1951, *The Preparation of Programs for an Electronic Digital Computer.* It was an immediate classic and foundational reference for anyone involved in the development of programming in the UK and abroad. The book was known by the names of its authors, often compressed to 'WWG'.

Framing EDSAC

Babbage predicted that without machine computation, the burden of calculation would stultify science. EDSAC 1 and its successor EDSAC 2 were early vindications of this prediction. Countless research projects exploited new-found high-speed electronic computing—statistics, astronomy, crystallography, molecular biology, geomagnetism, theoretical chemistry, mathematical physics, genetics, economic modelling, physiology, and radio astronomy. Three Nobel Prize winners number among those who used the Cambridge Mathematical Laboratory's new electronic computing facilities. For each, EDSAC had either accomplished otherwise impossible computational tasks or was of material aid in complex calculations.

In the general rush to make a usable stored program computer, Wilkes was first past the post. When asked how he had achieved this he said, others 'had committees to help them'. As the director of the Mathematical Laboratory, he had budgetary control and a set objective. He was spared the sapping embrace of bureaucracy. He described his *laissez-faire* authority as 'highly unusual'.

EDSAC marked a progression of the computer from a serious-minded curiosity, with great but yet ill-defined potential, to a usable machine through which computational possibilities could be explored.

As well as the new machines in Cambridge and Manchester, there was a third. Turing, one of the Bletchley team disbanded in autumn 1945, migrated to the National Physical Laboratory (NPL) in Teddington, south-west London. In February 1946 he produced an original complete design for a stored program computer, the ACE (see Box 1). A scaled-down version, Pilot ACE, ran in 1950. Turing's approach was unusual in relation to the developing design orthodoxies. Its instruction format was complicated, designed as it was for hardware economy, and programming was difficult. The pay-off, though, was speed.

Computers as products

The first-generation prototype computers spawned successors in Britain and America. Companies entered the market and computers became objects for sale. In the UK, Ferranti Ltd, Elliott Brothers, LEO Computers Ltd, English Electric, and others built and marketed computers from the early 1950s, vying with each other for a share in an up-and-coming new post-war industry.

The university prototype computers had been designed primarily for mathematical calculation and scientific computation. J. Lyons and Co., a catering company famous for its large network of tea shops, built their own computer for business applications. LEO I, closely modelled on EDSAC, was running in 1951. LEO (see Box 1) was used for payrolling, inventory, stock management, and scheduling for daily delivery of teashop supplies—40,000 different products to 150 teashops. We have here a shift from mathematical computation, the dominant preoccupation of computer-makers so far, to data management, business applications, and strategic planning.

While Britain stole a march on the Americans in the rush to realize EDVAC-type machines, the US scene was ramping up. The ENIAC/EDVAC group at the Moore School had dispersed. Von Neumann returned to the IAS, Princeton, taking with him key members of the Moore School team including Herman Goldstine, Adele Goldstine (his wife, herself a mathematician and programmer), and Arthur Burks. Eckert and Mauchly had defected to form their own computer company. Hampered by the loss of its stars, the Moore School struggled with EDVAC. The machine was redesigned, repeatedly modified, and was 'always threatening to work'. It ran its first program in October 1951 but was soon outclassed. It limped through until retired by the BRL at the end of 1962. The design concept of EDVAC was its illustrious distinction. Its service life was indifferent. Other machines were more successful embodiments of the EDVAC concept.

The most influential of the academic projects in the US was von Neumann's IAS machine. Von Neumann refined and developed his Moore School ideas, and with Herman Goldstine and Burks described these advances in a paper written in 1946. It was this, rather than the more celebrated 'First Draft', that served as a seminal reference for many seeking to build new machines. The IAS computer was completed in 1951 and publicly announced in June 1952. Government agencies and universities needed and wanted computers, and replicating the IAS machine seemed the most efficient way. A steady stream of visitors to the IAS departed with the requisite design drawings. Research institutes and universities secured grants to make copies. Many modified the master design, and variants of the essential machine proliferated. All had acronymic names (see Box 1).

In the US the commercialization of computers was spearheaded by entrepreneurs. Eckert and Mauchly were convinced of the commercial opportunities of building a general-purpose machine not specifically tailored to science and engineering but suitable for business use too. Here it would vie with established office practices using punched-card calculators and tabulators, a mature market dominated by IBM. Their computer was called UNIVAC—Universal Automatic Computer: 'Universal' to convey that it was general purpose, suitable for science as well as business; 'Automatic' to convey that its operation did not rely on office staff ferrying stacks of punched cards between machines and initiating their action. A major innovation was to use magnetic-tape storage for data instead of countless punched cards.

The financial history of the Eckert-Mauchly Computer Corporation (EMCC) was fraught. Advance-selling a half-million-dollar machine, unbuilt and unproven, to raise capital for development, was an ongoing struggle that ultimately failed. In 1950 Remington Rand, a large business machines manufacturer, bought EMCC. Eckert and Mauchly stayed on.

Election coup

In 1952, with only three systems installed, UNIVAC had a windfall. The US presidential election was contested by Dwight D. Eisenhower and Adlai Stevenson. CBS covered the event live on TV with a dummy UNIVAC in the studio as a prop, flashing lights and all. A working UNIVAC outside the studio was going to predict the result. With 7 per cent of the vote in, and against the forecasts of two major opinion polls and an army of pundits, UNIVAC predicted a landslide victory for Eisenhower. Election officials were disbelieving. They suppressed the result, and UNIVAC programmers tweaked the parameters to produce a more credibly close contest. After Eisenhower's landslide was publicly confirmed, UNIVAC's original prediction was revealed. The story created a media sensation that, in public perception, elevated the power of computers to mythical heights. As Hoover became synonymous with the vacuum cleaner, Kleenex with tissues, UNIVAC became the generic name for a computer.

History tells UNIVAC's tale as one of entrepreneurial and visionary drive. UNIVAC was a commercial success at least for a while. The first order was for the US Census Bureau. Corporations, industry, and government bought forty-six in all. Sceptics and doubters were won over. UNIVAC's sales were evidence of a sizeable market. Eckert and Mauchly had launched the US computer industry. Even IBM took note.

Chapter 5
The computer boom

The computer began to emerge from the workshop into the workplace. No longer one-of-a-kind machines attended by a technical priesthood, but a commodity product. No longer machines solely for scientific calculation, but machines for business and war. The centre-piece of the story from the early 1950s to the late 1970s is that of the corporate mainframe, and IBM's dominance.

IBM has its roots in the amalgamation, in 1911, of four companies. Two of the companies manufactured time-keeping instruments including employee time clocks. The third produced retail scales and food slicers. The fourth was the Tabulating Machine Company, founded by Herman Hollerith, whose card punches, tabulators, and sorters had transformed data processing for the 1890 US census. The merged company was called Computing-Tabulating-Recording Corporation (CTR). Thomas J. Watson Sr, CTR's president, renamed the company IBM in 1924. Hollerith's machines were the basis for IBM's punched-card data-processing product line.

Watson, a consummate salesman, autocratic and patriotic, offered IBM's services and manufacturing capacity to the US government during the war, capping profits from government contracts at 1.5 per cent. During the war, revenues from European sales doubled and IBM emerged as market leader in office automation and data processing. Its longstanding rival, Remington Rand, was a close second.

IBM stirs

Around the time UNIVAC hit the headlines by sensationally confounding the forecasted result of the 1952 US election, IBM had a large customer base locked into its office machines by familiarity and established practice. Rental income from its machines was the staple of its financial model. Management had recognized the potential of the new electronic technologies, but rather than launch a major redesign, it chose to incorporate electronics into existing products. The watchwords were continuity and evolutionary transition rather than anything sudden that would frighten the horses.

A computer as a specialized mathematical instrument for scientific calculation was a manageable threat. But a general-purpose computer that could be used for business, like UNIVAC, was a potential rival to be reckoned with. IBM's main business was based on electromechanical punched-card equipment for general office use, and they feared losing customers enthralled by the glamour and prospects of a new class of computers. UNIVAC's early sales were a wake-up call.

In response, Watson Sr poured resources into three machines—a scientific computer known as the Defence Calculator, and a business-oriented data-processing computer called the Tape Processing Machine, which used magnetic tape for bulk data storage and was intended to compete with UNIVAC. The third was a low-cost general-purpose computer that used a magnetic drum for its primary memory. The three machines were later known by their model numbers, 701, 702, and 650.

IBM avoided the word 'computer' in naming its equipment. Far from trumpeting the new as revolutionary and leveraging transformational promise as a plus, the company preferred to frame their new products as offering the security of continuity. Using familiar office equipment, customers could do what they had always done, but better and faster, and rest secure in IBM's fulsome service support. Continuity reassured against the unspoken fear of workplace transformation and the disruption that UNIVAC boded.

Watson Jr, Thomas J.'s eldest son, who ran the company from 1952 to 1971, described the frenzy of development activity as 'panic mode'. Nineteen 701s were installed, the first in December 1952, in IBM's offices in New York. The second went to the nuclear weapons labs at Los Alamos early in 1953. Eight went to aircraft companies. They were used for avionics, weapons design, space exploration, cryptanalysis, and military logistics. Customers extended their use to manage payroll and for financial and actuarial reports—UNIVAC terrain.

But it was the modest mass-produced 650 that was the star. Two thousand were delivered, the first in 1954, the last in 1962. In Watson Jr's words 'the 650 became computing's "Model T"'. IBM offered 650s to universities and colleges at 60 per cent discount with an obligation to establish courses in computing. The result was a generation of programmers and computer scientists familiar with IBM's products, and a recruitment pool with a fast track to employment on the mothership or in the field.

UNIVAC was a large monolithic construction assembled on the customer's site. IBM's computers were modular, consisting of separate units that could be flexibly configured depending on the specification of the system. The 650 had options for eleven separate units—card punch, card reader, magnetic-tape/disk controller, an accounting machine unit, tape and disk storage units, and tape-to-card punches. Crucially, each of the units could fit into a regular lift.

IBM's measures to counter the threat of UNIVAC succeeded. By 1955 the 700-series had outsold its feared competitor. By 1956, the year Watson Sr died, IBM was the largest and most profitable computer manufacturer. By 1960 it had over 70 per cent of the data-processing and computer market with $1.8 billion in annual sales and 104,000 employees. Game, set, and match. But not without some spirited and often despairing opposition.

Technology and the Cold War

Computers as commercial products have drawn the narrative away from technology towards business history. While computers were now viable products for sale, there remained residual problems of reliability. Vacuum-tube failures were one challenge, memory technologies another.

Early computers mostly used either mercury delay memory or Williams-tube memory. Both were complex technologies that could be skittish. A game-changer was magnetic-core memory developed by Jay W. Forrester at the Servomechanisms Laboratory at MIT.

In December 1944, with the war not yet ended, the US Bureau of Aeronautics asked MIT to assess the feasibility of a flight-training simulator to speed up pilot training and test new aerodynamic designs. A 'universal' simulator that could be programmed for different aircraft would spare the need for dedicated simulators, one for each aircraft type. Forrester was charged with the study. They got the go-ahead in 1945. The task was daunting. The simulator needed to react to the pilot's actions, drive the cockpit instrument panel to reflect ongoing flight metrics, simulate wind resistance, and continuously record flight data for later analysis—all this in real time. To provide a realistic training experience, the simulator's controller needed to react to pilot action credibly fast.

The default choice at the time was to use an analogue computer. This was soon abandoned as too complex, too imprecise, but more importantly, too slow. Forrester proposed a full-scale digital computer for the simulator's controller. The project was approved in March 1946 and named Whirlwind. Here he faced the same challenges as other post-war computer designers, principally the search for fast reliable memory. For real-time use the controller would have to run up to a hundred times faster than the scientific computers then envisaged, and nothing available, or in the pipeline, would do. Building the computer diverted resources from the simulator, which quietly faded as an expected deliverable.

Magnetic-core memory

Fast, reliable memory was a problem. Existing memory systems had drawbacks. In June 1949 Forrester experimented with a magnetic ceramic, and he had a prototype memory working at the end of 1951. Magnetic-core memory is made up of minute doughnut-shaped beads made of ferrite, a crystalline material with magnetic properties, arranged in a matrix. The beads are threaded with fine copper wires through which pulses of electric current magnetize or demagnetize the cores (see Figure 7).

Each core stored 1 bit (1 or 0), and electrical pulses allowed each core to be read from or written to. Forrester's team was not the only one experimenting with core memory, but Whirlwind was the first past the post with a working installation.

Core memory offered several advantages in addition to reliability. It is 'non-volatile'—the cores retain information in the absence of electrical power—unlike Williams-tube and delay-line memories, where the information vanished when the device was switched off. Core memory is also 'random access'—information in all cores is equally accessible, and retrieval from any core is equally fast. This is unlike delay-line memories, where information is in a serial string, and you have to wait for what you want to stream past.

7. Magnetic-core memory showing matrix of 'doughnut' beads (1972). Circular inset shows magnified detail.

In summer 1953 Whirlwind's problematic cathode-ray-tube memory was replaced with magnetic-core memory, making it the fastest and most reliable computer to date. ENIAC's memory was replaced with core memory at about the same time and within five years core memory was the memory of choice across the industry. Commentators have argued that the benefits to computing of finally having fast, reliable memory, a spin-off of the Whirlwind project, more than justified the wincing cost of the whole project.

In an early 1970s interview Forrester himself said that Whirlwind 'probably contained more features that survived into today's computers than any other machine of its time'. Core memory is perhaps the most well-known. Less known is the use of a light gun that allowed an operator to interact with a cathode-ray-tube display screen, the first human–machine interaction of this kind.

Reliability

Vacuum-tube failures were a prevailing concern. Tubes were thought to have a life of about 500 hours. A machine with 20,000 tubes would be expected to fail every few minutes. Forrester's team found solutions. One was for the computer itself to run internal test programs to assess the condition of the tubes by detecting marginal deterioration and so anticipate failures. They also found that tubes were failing because of an effect produced by silicon in the metal from which the cathode was made. Removing silicon from the manufacturing process increased tube life from 500 to 500,000 hours. Combining this improvement with marginal checking gave an effective life of five million hours. These and other measures, such as close control of component tolerances and screening out extraneous interference, resulted in weeks of continuous error-free operation.

Forrester remarked that it was a running battle to sustain funding to continue development. What he said next is a telling reminder of the danger of uncritically projecting our own attitudes onto the circumstances of an earlier time. 'People were unconvinced of the idea that a computer could substitute experience and judgement and decision making... That was particularly true in the late 1940s when no one had ever seen it happen'.

Air defence

The Cold War saved Whirlwind. The Soviet Union exploded an atom bomb on 29 August 1949 and had developed long-range bombers capable of penetrating to the US. The existing air defence system was a patchy collection of radar stations on the east and west coasts, leaving approaches from the north exposed. The US Air Force's response to the perceived threat of Soviet airborne attack was ambitious and robust—to develop a comprehensive computerized national early-warning air defence system. Cost no object. No expense spared.

The system was called SAGE (Box 1). Defence was based on interception by fighters of incoming bombers. SAGE had to detect, identify, and track incoming aircraft using input from radar, and compute an interception course for each to allow a controller to guide a fighter, by radio, to the approaching target. Alterations of bomber flightpath required continuous updates and real-time computation of directions for the responding fighter. The calculation of intercept trajectories was time-critical. Whirlwind, the only real-time digital electronic computer, was the prototype for SAGE. Forrester and his team led the new development.

Production versions of Whirlwind were built by IBM, selected as the prime contractor in 1952, after rigorous evaluation of four contenders, one of which was its old rival, Remington Rand. The computer was called the AN/FSQ-7 (Q-7 for short). IBM's contract was for fifty-six SAGE computers, a pair for each installation, at a cost of a half-billion dollars.

Uninterrupted vigilance was imperative, and each station had sleeping quarters for operators. Consoles had cigarette lighters and ashtrays built in. Each installation duplicated essential hardware, one running live, the other on hot standby with provision for switchover in the event of failure. Combined reliability measures resulted in computer down-times of no more than a few hours per year.

SAGE installations were housed in twenty-three Direction Centres, each one of which covered one sector, several thousand miles square. Collectively the sectors quilted the country, and coverage overlapped into the sea. Each Centre had a Q-7 weighing 250 tons with a total of 60,000 tubes—49,000 for the computers, the largest yet built. A single SAGE installation could run fifty display monitors and track up to 400 aircraft. The overall system could co-ordinate and integrate information from up to a hundred radars—ground- and ship-based, on-board missile and aircraft radar, observation stations, and other SAGE Centres in the system—to produce a unified image of the airspace.

The first Centre was operational in July 1958. The last was shut down in January 1984 after an unbroken service life of twenty-six years. Performance took priority over cost, which ran at an estimated $8–12 billion.

Innovation spinoffs

Whirlwind and SAGE are amongst the most innovative and influential systems in computing history and the value of technological spin-offs from SAGE into civilian computing is incalculable. New technologies included magnetic-core memory, interactive screen display, printed circuits, mass storage devices, real-time operating software, duplex standby, digital communication over standard telephone lines, time sharing, programming disciplines, and circuit assembly using automatic soldering on double-sided printed circuit boards. These techniques and technologies laid much of the foundation for what followed.

One story of Whirlwind and SAGE is that of technological innovation. Another is of the migration of people and technology to the civilian computing industry. While IBM was the principal contractor, it was far from alone. The project enlisted the technical resources of major corporations—Burroughs, Bell Labs, MIT's Lincoln Laboratories, Western Electric, Rand Corporation, and countless smaller contractors and manufacturers. The upshot was the effective diffusion of innovation and expertise from the military into civilian environments, which gave US companies a head start in computing, networking, and communications. IBM was a privileged beneficiary, By 1958 over 7,000 staff, some 20% of its workforce, were working on the project, which generated 80 per cent of its revenue between 1952 and 1955. As prime contractor IBM secured and retained a lead in real-time systems, memory, mass storage, and processor technologies that it successfully spun off into commercial products.

A parallel frontier was software, no less demanding in scale and capability though receiving less attention by historians. SAGE trained and employed half the estimated total US programming workforce—over 700 programmers who spent some 1,800 programmer years to develop SAGE's software systems. The outcome for commercial computing was a highly trained specialized cohort of programmers with state-of-the-art expertise, who dispersed into commercial computing to develop new products, systems, and applications.

Enormous computers, vigilant operators staring at flickering screens, darkened rooms in the grim windowless four-storey concrete blockhouses of the SAGE installations are part of Cold War legend (see Figure 8).

SAGE computers, control consoles, and screens were coveted props in many movies, not least the darkly absurdist *Dr Strangelove*, which agitated concern about the wisdom of relegating the fate of humanity to machines and algorithms, now that these had become powerful enough to threaten survival.

In the four-thread scheme, SAGE made radical advances in three: automatic computing (plotting intercept trajectories in real time), communications (radar and interstation digital communications, operator communications using interactive screens), and information management (real-time screen-based depiction of airspace). SAGE integrated the developing technologies of three separate threads.

8. Console of SAGE early-warning air defence system deployed in the US during the Cold War against the threat of Soviet airborne attack.

Two sectors of commercial computing were quick to exploit and deploy new expertise and technologies: banks and airlines.

The cheque crisis

In parallel with the crisis of perceived airborne attack, there was a banking crisis. Or rather, more of an information management crisis not unlike that of the 1890 US census. In 1952 eight billion cheques were written in the US, doubling from four billion in less than a decade. Predictions put the number at fourteen billion by 1960. Each of the twenty-eight million cheques written every business day was processed by hand, mostly by young female bookkeepers, few of whom tolerated the drudgery of it for long. Staff retention was one issue and there was anyway only a finite pool of young bookkeepers from which to recruit. Keeping pace with rising volumes from new business was increasingly fraught. What banks needed was ERMA (Electronic Recording Machine—Accounting), the uncomplaining employee.

Bank of America, the largest bank worldwide in the 1950s, saw the paperwork crisis as an obstacle to growth and sought rescue in automation. The bank's then vice president, one S. Clark Beise, established a collaboration with the Stanford Research Institute (SRI), Menlo Park, California, to develop ERMA. Issuing punched cards as cheques would have provided a solution using existing office machinery for counting, sorting, and bookkeeping. But the Bank held that the cheque was an important 'emotional link' to the customer and that the conventional format and appearance was sacrosanct. A punched card as a cheque just wouldn't do. Instead SRI developed magnetic ink to print a machine-readable account number on regular cheques. Up till then customers were identified alphabetically by name.

A special machine-readable font was designed to maximize discrimination between characters. To aid reliable character recognition, printed numbers had strange bulges, distortions, and odd thickenings but were nonetheless readable by humans. The MICR (Magnetic Ink Character Recognition) font, agreed in 1959, became the banking standard and remains in use. The chunky funky font was seized on by the advertising industry as the typographic symbol of automation and computer culture.

General Electric (GE) got the contract to make thirty-six production versions of the ERMA prototype with a requirement that each installation be capable of processing 55,000 transactions a day. In 1955, GE was the leading US electronics firm, with $3 billion sales and 200,000 staff compared to IBM's $461 million and 46,000 staff. GE was nonetheless a curious choice—it had no computing department and was anyway against entering the business-machines market, focusing instead on one-off contracts with military agencies. The first ERMA system was installed in December 1958 at the San José branch, and ERMA was publicly unveiled the following year. Ronald Reagan, actor and TV presenter (and future US president), hosted the inauguration.

By the time the Bank of America installed the last ERMA, in 1961, the system was servicing 2.3 million checking accounts at over 200 branches, and the bank stopped hiring bookkeepers. Fears about job losses fuelled the debate in Europe and the US about the threats of automation to traditional employment.

Credit cards

ERMA prompted the introduction of the first credit card. Diners Club and American Express had launched credit cards usable in specific sectors—restaurants, travel, and entertainment. But the balance was payable in full each month. BankAmericard was the first to offer a general-use card with deferred payment allowing card-holders to pay down an outstanding balance over time. Bank of America introduced the card in 1958 and licensed its credit card operation to other banks. In 1970, issuing banks formed an interbank card association, National BankAmericard Incorporated. In 1976, BankAmericard became Visa.

The significance of ERMA was not its computational capabilities. These were undemanding—bookkeeping arithmetic. Nor was it primarily communications—the installations were at independent regional centres, and cheques were processed in batches using a messenger service for delivery. In the four-thread scheme, ERMA, which automated high-throughput cheque processing and account transactions, is squarely part of information management.

Airline reservation

In the early 1950s airlines began operating commercial jet passenger services. With the rise in passenger numbers the booking process was under siege. Like the cheque-clearing crisis and the 1890 US census, the existing airline seat-reservation system, managed manually, was reaching its data-processing limits. Airlines ran flights at below capacity as a cushion against an inflexible and inefficient reservations system. The high capital cost of the new passenger jets, increasing competition, and greater price sensitivity, pressured airlines to increase profitability by maximizing the number of occupied seats. The solution was a direct beneficiary of the SAGE project that had pioneered integrated real-time data processing.

By 1953 American Airlines acknowledged the increasingly chaotic system in which seat availability at any one time was often impossible to establish with any certainty, and a booking procedure that involved a tortuous chain of clerks and operators in a multi-stage manual process. Following a chance meeting in 1953 between the Airline's president and a senior IBM salesman on a flight from Los Angeles to New York, IBM and American Airlines collaborated in a project to computerize the reservation system. Within IBM the project started as part of an existing programme called SABER, a tortured acronym that later gave way to SABRE, not an acronym but a name suggesting incisiveness and swashbuckling panache. After SAGE, IBM was uniquely placed to develop an integrated real-time computerized information and communications system, requirements the two projects shared.

SABRE was the largest commercial computerization project undertaken till that time, and was fully operational in December 1965, after over a decade of research and development. The system integrated reservations, ticketing, check-in, and passenger records. Seat inventory was centralized, updated in real time, and information was networked through over 10,000 miles of leased lines to over 1,000 agents in fifty US cities. As with SAGE, the IBM mainframe running the system was duplicated. One was on permanent standby with automatic switchover in the event of failure. The system could handle some ten million reservations a year, 20,000 ticket sales and 85,000 telephone calls a day. Transaction times were reduced from hours to seconds. System cost: $30 million.

With SAGE the imperatives were military. With SABRE, they were commercial. Other airlines couldn't afford to lose competitive advantage. Delta, Pan American, and Eastern Airlines followed, with IBM the willing systems supplier. Other providers entered the market and by the early 1970s all major carriers worldwide had integrated real-time reservation and networked communications systems. In the four-thread scheme SABRE, like SAGE, saw the convergence of three threads—information management, communications, and automatic computing.

IBM and the seven dwarves

Remington Rand and the UNIVAC computer were far from IBM's sole competitors. Companies with electronics and manufacturing expertise found custom in scientific computing. But when the computer became a high-volume business machine, only the largest could contemplate taking on IBM with its legendary sales, marketing, and customer service operations, established customer base, formidable research and development, and in-house manufacturing resources. By the late 1950s, the computer industry was made up of IBM and a clutch of hopefuls dubbed by the press 'the seven dwarves'—GE, Burroughs, Sperry Rand (UNIVAC), National Cash Register (NCR), Radio Corporation of America (RCA), Control Data Corporation (CDC), and Honeywell.

Two products consolidated IBM's dominance and furthered the shakeout: the 1401 computer, announced in October 1959, in which transistors replaced vacuum tubes, and core memory replaced magnetic-drum storage. And then the System/360, announced on 7 April 1964, that transformed the industry.

Computers proliferated as organizations computerized punched-card procedures. The 1401 was nonetheless unexpectedly successful. The feature that boosted its popularity was the high-speed printer capable of 600 lines per minute, four times that of the most popular accounting machine at the time. Sales tipped 12,000 between 1959 and 1971, when IBM retired the line. This was way beyond projected sales. To signal the new, the 1401 came in rectangular light-blue cabinets. The blue livery became an increasingly widespread feature of office environments as companies upgraded and new clients bought in. IBM acquired the moniker 'Big Blue' echoing the sinisterly ubiquitous Orwellian dominance of the corporate mainframe and centralized control.

In the 1960s IBM was supplying and supporting seven different computers, each with its own marketing personnel, production line, software, and specialized hardware. Each model of peripheral required a different controller to attach to each of the different processors. Siloed diversity was costly. But the major problem was the lack of software compatibility across the model range. Software was written for a specific processor. If you upgraded to a larger computer, or downgraded to a smaller one, software had to be rewritten, an expensive, disruptive, and time-consuming duplication.

System/360

IBM's solution was radical: to develop from scratch a compatible family of computers for which the same software would run on any model. The new system was called System/360, so named to suggest all-round 360-degree capability. IBM had your back. The new product line would make all IBM's existing machines obsolete and redefine the industry. A *Fortune* journalist called the project 'the $5 billion gamble', a bet on the company's future. Watson Jr and his second-in-command, Vincent Learson, approved the programme on 4 January 1961, against internal caution and indifferent support from top management.

IBM was bumped into announcing the System/360 while still discussing a roll-out strategy. Its hand was forced by Honeywell which announced, in December 1963, a computer that was software compatible with IBM's popular 1401 but with significant price-performance advantage. Honeywell landed 400 orders in the first week and IBM feared losing new orders as well as existing 1401 customers to their no-nonsense new competitor. IBM countered by launching the entire System/360 product line in one hit, well ahead of what was planned. On 7 April 1964, 100,000 customers and journalists met in 165 US cities to hear the wonders of the new system. IBM held press events in fourteen countries. Unveiled were six new computers with a performance range of about 50:1 between the largest and smallest, forty-four peripherals, and the promise of software that was compatible across the range. Equipment still in development was displayed as wooden mock-ups.

Industry and users were stunned by the scale, ambition, and boldness of the scheme. IBM took over 110,000 orders in the first month from customers worldwide. Production was now the challenge as IBM struggled to deliver. It opened new plants and hired over 70,000 new staff.

The 1960s saw the prolific expansion of computers into business practice. In 1973 the number of computers worldwide tipped 100,000 for the first time. IBM's position seemed unassailable, with System/360 technology the basis of IBM's product line into the 1990s. Its mainframe architectures and standards are its lasting legacy. By the mid-1970s the US computer industry consisted of two players: IBM on the one hand with about 70 per cent of the market, and everyone else combined. The gamble paid off.

Of the seven dwarves, five survived with a variety of strategies to secure small percentages of the remaining markets. GE and RCA withdrew from the fray leaving Burroughs, UNIVAC, NCR, Control Data, and Honeywell, known collectively by what else but an acronym—the BUNCH.

Chapter 6
Revolution

Historians of technology are wary of 'revolutions'. The word smacks of journalistic hyperbole. The idea of revolution trades for shock value on the concept of sudden and seemingly unexpected change. Yet the success of computers as product, their capability and performance, cannot be understood without some sense of what those who use such language have in mind when they talk of 'the micro-electronics revolution' or 'the computer revolution'.

Solid-state electronics

Advances in electronics, specifically semiconductor electronics, transformed computer design. First transistors, then the integrated circuit (IC) and then the microprocessor—the 'computer on a chip'.

The transistor was invented in 1947 at Bell Laboratories, after a concerted effort by a specially formed group of mainly theoretical physicists headed up by William Shockley, himself a physicist. The name 'transistor' was coined at Bell Labs and is thought to be a compaction of 'transresistance' and a suffix '-istor' used for other generic components as in 'varistor' and 'thermistor', though origin accounts vary. Bell's interest was to find a durable, reliable replacement for vacuum tubes and relays. Transistors have no heater filaments and need neither the power nor warm-up times of vacuum tubes. They are power-efficient, small, and physically robust. Immunity to shock had appeal for military applications where portability in hostile environments was at a premium, as were size and power-economy for use in missiles and aircraft.

The implications for computing were not immediate. It took over a decade for transistors to find their way into computers. The first transistorized products for public consumption were hearing aids (1953) and the portable transistor radio (1954). Military, then commercial computers turned from vacuum tubes to transistors from the mid-1950s, and early military demand did much to fund development and allow cost-benefits from the economies of scale to make ICs a commercial proposition.

Transistors are made from semiconductor material, so called because its electrical properties lie somewhere between that of good conductors of electricity (like copper and many metals) on one hand, and insulators on the other (such as plastic or ceramics) which are very poor conductors. First-generation transistors used germanium, a naturally occurring crystalline material that was selectively 'doped' by the controlled introduction of impurities to alter its electrical properties and allow it to act as an amplifier or a switch. Silicon, also naturally occurring, duly replaced germanium as the preferred substrate for transistors and for the ICs that followed. The behaviour of semiconductors is the theoretical domain of solid-state physics, which was at first pressed to describe the new conductive phenomena. So prolific is silicon in electronic devices, and so far-reaching are the consequences of semiconductor technologies, our Silicon Age is said to have superseded the Iron Age.

A single semiconductor transistor was sealed in a small package, sometimes metal, sometimes plastic, about a centimetre or less at its widest, with three wire leads protruding. Each transistor was an individual, separate device. Hundreds and thousands of transistors were mounted on boards with other components and wired together to make the complex logic circuits that are the foundational parts of a computer. Vacuum-tube computers are first-generation machines. Discrete-component transistor computers are second generation, fairly short-lived in design terms though not necessarily in terms of service life. It is third-generation devices, based on the chip, or integrated circuit, that lend meaning to the language of 'revolution'.

Integrated circuits

The IC placed all the components of logic circuits—transistors, passive components such as resistors and capacitors, and interconnections—on a single tiny block of silicon in a way that integrated these parts into the molecular structure of the chip. The integration of circuits into a solid block of silicon had radical consequences for reliability, reduced size and weight, efficiency, and eliminating the need for external wiring to connect together large numbers of separate components. Circuit components were no longer 'assembled' but 'grown' into the chip. External wiring, with the attendant risk of error, was replaced by printing—a photolithographic process used in the fabrication of ICs that defined the detail of the components in the monolithic structure.

To allow external connection, the minute silicon block was typically encased in a black plastic rectangular package to protect it, and give it enough perimeter for connection pins. ICs in the 1970s typically looked like regimented black caterpillars with two rows of shiny metal legs (see Figure 9).

There were dozens of initiatives worldwide from the 1940s to the 1960s by scientists and engineers to produce micro-circuits. The developmental stages, and the diversity of alternatives, makes for a complex tale in which priority and inventive credit remain contested. Unreconstructed histories privilege heroic inventors, visionary flashes of insight, isolated discovery. Such narratives reduce histories to manageable accounts essentially by

9. Integrated circuits (ICs) or 'chips', typical of the 1970s.

omission—of context, cross-influence, compound agency, nuance, and the unexplained. The most widespread, if simplified, narrative about the invention of the IC is one that largely succumbs to these tropes. The two-inventor account is the one that mainly prevails. It confers joint honours on Jack Kilby, an engineer at Texas Instruments, Dallas, and Robert Noyce at Fairchild Semiconductor, in Mountain View, California.

Kilby was awarded the Nobel Prize for Physics in 2000 for his part in the invention of the IC. Noyce had died in 1990, and Nobel Prizes cannot be awarded posthumously. Their foundational works were conducted in 1958 and 1959, and were followed, inevitably, by patent disputes about priority.

The first markets for ICs were, yet again, government and the military. In the mid-1960s the US Airforce re-engineered the Minuteman intercontinental ballistic missile. The supply of the 4,000 ICs a week for Minuteman II funded volume production for

ICs. John F. Kennedy's declared goal in 1961 to put a man on the Moon by the end of the decade created new urgency. Contracts followed with Fairchild Semiconductor from the National Aeronautics and Space Administration (NASA) for the guidance computer for the Apollo spacecraft. The Apollo programme successfully landed the first astronauts on the Moon in 1969, and five more times before the end of 1972. Each guidance computer contained about 5,000 ICs, and seventy-five systems were built. In the mid-1960s most of all ICs in existence were in Minuteman guidance systems and in Apollo spacecraft. Production for NASA and the military laid the technical and economic foundation for commercial use.

Moore's Law

Semiconductor technology transferred electronic circuitry from external hardwired connections between separate components to the internal domain of molecules. Refinement of the fabrication process consistently increased achievable component density. That the rate of improvement was sustained with such constancy is a remarkable anomaly of this technology.

In 1965, Gordon Moore, director of research and development at Fairchild, observed that the number of components on a chip had doubled each year for the past four years, and speculatively predicted that this rate would hold for a decade. In 1975 he revised his projection from doubling every year, to doubling every two years, and this prediction became known as Moore's Law. The 'Law' is not a physical law in the sense of the Second Law of Thermodynamics, or the Conservation of Energy. It is not an expression of the impossibility of deviating from the compelling inevitability of a naturally ordained outcome. It is a rate of improvement extrapolated from an observed historical trend in industrial production. Maintaining Moore's Law became an aspirational driver for IC fabrication. What is remarkable is the consistency with which the industry succeeded in sustaining Moore's Law, year in, year out, to the present time.

Revolution

How credible is the notion of revolution in relation to computers? Is it historically responsible to talk in this way? Metrics suggest an answer. Crossing the Atlantic by ship in the late 1940s took about five days. Crossing by supersonic Concorde in 1996 took just under three hours, and we would probably agree that this represents a revolution of sorts in passenger transport. The reduction in time is by a factor of forty. An off-the-shelf laptop now runs faster than the 1949 Cambridge University EDSAC computer by a factor of 6,000.

Speed is one index. Size another. If we compare the reduction in physical size of the active elements of computers over the same seven decades, we find ourselves comparing a vacuum tube in EDSAC measuring 7 centimetres tall, with the submicron geometry of a modern silicon chip. One micron (a millionth of a metre) is the length your fingernail grows in about ten minutes.

Fairchild's first IC, announced in 1961, had four transistors and five resistors. By 2020 the number of transistors in a commercially available IC was in excess of thirty billion. At current densities the reduction in size from vacuum tube to transistor-on-a-chip is by a factor of several million. Compared to the 'revolution' in passenger transport, changes in computer performance over the same period signal something technologically unprecedented.

Minicomputers

In the mid-1960s there was a new arrival on the block—the minicomputer. The metaphor might seem glib, but it captures the sense of an outsider, new to a settled neighbourhood. The newcomer did not threaten the leased-mainframe data-centre model of corporate computing, but opened up new environments, applications, and class of user. The minicomputer represents a distinct phase in the development of manageably small, lower cost computers that allowed, and encouraged, a more direct relationship between user and machine.

The central protagonist in the minicomputer narrative is DEC (see Box 1) founded by Kenneth ('Ken') Olsen and Harland Anderson in 1957. Olsen, an MIT engineering graduate, had worked on magnetic-core memory for Whirlwind and at MIT's Lincoln Laboratory, a government-funded research centre where he led the development of an advanced early transistorized computer. So this new class of machine had its roots in the military electronics industry rather than in corporate computing.

The PDP-8

DEC launched the quintessential minicomputer in 1965, the rugged, highly successful PDP-8, a germanium transistorized machine with magnetic-core memory. Computers at the time were large expensive mainframes. The PDP-8 was the size of an undercounter domestic fridge and cost $18,000, a small fraction of the cost of a mainframe, and a price unheard of for a computer. Its capabilities were nowhere near that of a mainframe but it had features that appealed to a new class of user. PDP stood for 'Programmed Data Processor'. Olsen pointedly avoided the word 'computer' so as to lower sticker-price expectations, which for mainframe computers were in the order of $1 million. DEC was redefining what a computer was understood to be.

Olsen's company was the cultural antidote to IBM. No prestigious modern premises. Instead, a barely refurbished old woollen mill building for the Massachusetts headquarters. No plush chairs in swish reception areas with glossy magazines for waiting visitors. Small design teams. Disdain for marketing. Skeleton sales staff. Negligible customer application software. It was a stripped down unconventional spartan operation, free of the conventions of corporate computing. No-frills was a badge of pride. Computing had loosened its tie.

Later versions of the PDP-8 took advantage of higher component densities offered by ICs, and the purchase price kept dropping. Though still priced above ordinary personal affordability, the cost of a machine was, for the first time, within departmental budgets and this opened the floodgates. Individual users, divisions of corporations, research laboratories, hospitals, universities, eager for access to computing facilities, found funding for computer purchases within departmental budgets. Minicomputers had no need for large air-conditioned rooms or armies of technicians. The stakes were now lower, and so was the level of management scrutiny and control, and, critically, purchase approval. PDP-8s were released in various versions, the last in 1979. Some 50,000 PDP-8s were sold in all.

DEC doled out technical manuals, circuit diagrams, and specifications, material till then regarded as proprietary and to be protected, but now just given away. Users programmed their own specialized applications. This was encouraged by DEC. With information about the machine's circuits freely available, users were able to interface the computer with external systems. This created new opportunities for real-time data capture, and PDP-8s found widespread use in laboratory instrumentation, factory process control, medical research, machine-tool control, science labs, and countless applications in experimental research. A coveted mini was commonplace in many university science and engineering departments. Users, kept at bay in mainframe data centres, could now experience hands-on computing. Many had near one-to-one access to their computers and many of those developed a sense of personal ownership of their machines. Seeds of the personal computer—outright purchase and individual control.

The minicomputer class also created a market for embedded computers—PDPs as a component in larger special-solution systems, made, programmed, rebranded, and sold by third parties—the so-called OEM (Original Equipment Manufacturer) arrangement between, say, DEC as the computer supplier, and a third-party developer.

The company prospered, and the sector grew. New as well as established companies entered the market—Data General, Scientific Data Systems, Hewlett-Packard, and Honeywell were among about a hundred companies, new and established, that offered minicomputers between 1968 and 1972. By 1974 there were 150,000 minicomputers in service. In 1988 DEC was the world's second largest computer company next to IBM.

Minicomputers as a class died in the mid-1990s but their legacy is deep and lasting. DEC no longer exists. It was bought by Compaq in 1998. Compaq and Hewlett-Packard merged in 2001. History puzzles over DEC's demise, discomfited by its fate.

A problem of recency

To take the tale further lures us beyond the settled accounts of canonical history. The historians Martin Campbell-Kelly and William Aspray offer the telling case of domestic radio, noting that it took about five decades to properly historicize events. The interplay of commercial, technical, and cultural factors, the critical role of amateur radio hams, enthusiasts and hobbyists, and the institutionalization of public broadcasting, took reflection and analysis only possible with distance and time.

When it comes to events of the last decades—the personal computer, internet, worldwide web, smartphone, social media—we are still too close for balanced perspectives. The historian Paul Ceruzzi describes the trajectory of the personal computer, for example, as 'difficult to describe as rational'. The narrative was first framed largely by journalists from press releases, product launches, PR, advertising, corporate gossip, promotional trade literature, and interviews. The tale is constructed using recognizable tabloid standbys—isolated inventors, commercial rivalry, visionary enterprise, undreamt of success, improbable fortunes, opportunity, characterful actors. Writing to astound distorts history. Historians have done much to provide correctives, but the story of the personal computer is still playing out. Unlike the minicomputer, the personal computer as a class is far from dead. It seems that we cannot wait for history. So here are the headlines.

The microprocessor

It is the unforgotten desktop calculator that prompted the technological advance that is the engine of the personal computer. A Japanese calculator manufacturer, Busicom, was stung by the introduction, in 1969, of a new electronic calculator by Sharp, a Japanese electronics company. The new calculator, using advances in large-scale chip integration, reduced some 200 components to four custom ICs. Busicom sought to replicate this success and approached the US semiconductor company, Intel, to supply the ICs. Intel, founded by Robert Noyce and Gordon Moore, had only just been incorporated. At the time, making custom ICs was not Intel's preferred business, specializing as it did in semiconductor memory as a replacement for magnetic-core memory.

Marcian ('Ted') Hoff, Stanford PhD and Intel engineer, persuaded Busicom to abandon the custom chip approach in favour of a general-purpose IC that could be programmed to behave, amongst other things, like a calculator. The generic chip, a microprocessor, with three other ICs in the chip set, provided the essential logical blocks of a general-purpose von Neumann computer. The main chip was the now legendary Intel 4004 microprocessor containing 2,300 transistors, advertised in 1971 as 'a microprogrammable computer on a chip' heralding 'a new era of integrated electronics'. With a general-purpose device of this kind, altering what it did shifted from soldering to programming.

Ted Hoff is celebrated for the invention of the microprocessor. Credit for the invention is still debated. Hoff was not alone. There were contributions from Frederico Faggin of Intel, who led the engineering work, Intel engineer Stanley Mazor, and Masatoshi Shima, a Busicom engineer. Intel was not alone in proposing the microprocessor, which was conceived independently by others. Historians niggle about simplified accounts. Someone has to maintain standards.

The personal computer

The microprocessor is the engine of the personal computer but its embrace was not immediate. It took till 1977, the year the Apple II was announced, for the personal computer to be publicly acknowledged as an identifiable class. There are several cultural, commercial, and technological movements that combined to realize what seemed at the time an improbable success. A fertile context was provided by hobbyists, with their love of electronic construction kits and a passion for computing. The Homebrew Computer Club, first convened in March 1975 in a garage in Menlo Park, California, gave organizational focus to amateur interest. The Club provided a forum for computer hobbyists, acted as a clearing house for news, innovation, and expertise, and it published an influential newsletter. Meetings soon migrated from the garage to an auditorium at the Stanford Linear Accelerator Centre (SLAC).

Other protagonists were 'computer liberation' idealists, part of the countercultural legacy of the anti-establishment 1960s. 'Computers for all' was an ideological rallying cry. Personal empowerment through technology. Stewart Brand, who created *The Whole Earth Catalog*, first published in 1968, was a figurehead for the movement that advocated technology as a tool for personal liberty, fulfilment, and self-sufficiency. Computers were cool.

The Altair 8800 lit the touchpaper. It was a mail-order computer kit to be assembled at home. It cost $397 and was supplied by a small kit-supplier, Micro Instrumentation Telemetry Systems (MITS) in Albuquerque, New Mexico. MITS advertised the Altair in the January 1975 issue of *Popular Electronics* magazine as the 'World's First Minicomputer Kit to Rival Commercial Models'. In its native state the basic machine, once assembled, didn't do much. No keyboard, no screen. No software. Programming required commitment and time—flipping switches on a front panel to enter instructions in binary. There were extra slots for add-on boards that allowed users to extend capacity and function—memory expansion, interfacing with a Teletype, an audio cassette player/recorder (for data storage), and ports for peripherals. Many of the add-ons were made and sold by hobbyist-entrepreneurs and represented a modestly lucrative aftersales market.

For most of those who bought the Altair it was less a case of what you could do with the machine than the opportunity to get hands-on experience with a computer. The Altair was about a promise of what might be. It swept the board for the first half of 1975 with over $1 million in orders in the first three months.

Microsoft and Apple

The launch of the Altair attracted the attention of two friends, both avid programmers, William ('Bill') Gates, a Harvard freshman, and Paul Allen, a young programmer at Honeywell. Gates called Ed Roberts at MITS, one of the three designers of the Altair, and offered software that would allow users to program the Altair in BASIC, a popular, easy-to-use high-level language that became the default programming language for home computers. They wrote Altair BASIC in six weeks and MITS agreed to distribute it. Allen joined MITS as 'software director' and Gates quit Harvard and joined Allen in Albuquerque a few months later. The pair co-founded Microsoft (initially Micro-Soft) on 4 April 1975. They were now in the personal computer software business, where they had wanted to be.

The commercial potential of the personal computer market was unmistakable and by 1977 there were at least thirty companies in a competitive free-for-all. One of these was Apple.

Steve Jobs and Steve Wozniak (informally 'Woz'), two young enterprising computer hobbyists, founded Apple in April 1976. There was a third founder, Ronald Wayne, who, for $800, sold back to Jobs and Woz his 10 per cent stake in the company shortly after Apple was founded. Wayne had no regrets. He was, by his own account, 'too old' for the roller-coaster fortunes that he foresaw and did not wish stress to earn him the distinction of being 'the richest man in the cemetery'.

Woz was a Homebrew enthusiast from the start and a gifted electronics designer. He built a single-board computer using the cheapest microprocessor on the market at the time, the 6502 by the company MOS Technology. Woz's motives were not commercial. He just wanted 'to go down to the club and show off'. The naked single-board computer was called the Apple I. Jobs wanted to go into business and secured an order for one hundred Apple I boards from Byte Shop for $500 each, which gave sufficient financial credibility to found Apple Computer Company. Operating out of Jobs' parents' garage, they sold, overall, about 200 boards.

While Woz developed his next machine, the Apple II, Jobs secured investment capital, savvy management, and the services of a prominent PR company. This groundwork lifted the company above the mass of start-up hopefuls. Jobs and Woz launched the Apple II at the West Coast Computer Faire, an annual computer industry conference and exposition intended to popularize home computing. The event was held in San Francisco in April 1977. The Apple II was sold pre-assembled, housed in an elegant, distinctive cream-white moulded plastic case that could be taken home and plugged in to use. No soldering or arcane know-how needed. There were several expansion slots for cards that allowed third-party vendors to provide compatible devices—memory boards, disk controllers, network cards, modems for communications, and various interfaces. It had colour graphics capability, prized by computer games enthusiasts.

Woz published the design in *Byte* magazine in May 1977, an implicit invitation for crowdsourcing innovative applications and plug-ins. One prized card was the Z-80 SoftCard that allowed the Apple II to run the widely used CP/M operating system that gave ready-made access to a range of applications software including word processing, spreadsheets, and data-base management—the principal applications for small businesses. Software was critical to usefulness and appeal, and countless firms sprang up to supply games, and educational and business software applications.

Competition was fierce. The three main rival machines, all released in 1977, were the Apple II, Radio Shack's TRS-80 (called by its detractors the 'Trash-80'), and the Commodore PET. Apple boomed and became the fastest growing company in US history. The Apple II series sold some six million in the sixteen years of its production between 1977 and 1993. It is the Apple II that has come to symbolize the coming of age of the personal computer as a consumer product.

Untold histories

The genesis of the personal computer is an American tale. The Homebrew Club and the culture of the casual are part of the folklore of the period. The 'garage-to-multinational' story of Apple has become a parable of our times, celebrating entrepreneurial success, youthful cheek, and the American dream.

Histories of computing privilege success and this leads to the neglect of related narratives. The Soviet Union had a parallel 'Homebrew' drive but with a bleaker outcome. Four hip students at the Institute of Informatic Systems, Novosibirsk, Siberia, were frustrated in the mid-1980s at the lack of personal computers for home use. They designed, for themselves and a group of friends and colleagues, a personal workstation, Kronos. Without access to venture capital and open markets, only modest numbers were built, and the initiative, technically successful, did not ignite an indigenous personal computer industry as had the Altair and the Apple in California. To satisfy demand, the Soviet home market tracked the West with its Apple-compatible AGAT, and by reverse-engineering (copying) the Sinclair ZX Spectrum and other Sinclair home computers released in the UK in the early 1980s.

The IBM PC

Tinkering at home with kit computers was no threat to the computer industry. But when, by 1980, it was clear that the personal computer had a clear role as a business machine, it was trespassing on reserved territory. IBM, aloof and watchful till then, took the plunge and entered the personal computer market. In the interests of speed, and with unexpected suppleness, it created a fast-track programme for the new project. It abandoned its bureaucratic development process and its tradition of manufacturing all components in house.

In August 1981, after no more than a year, the company launched the IBM Personal Computer (PC) at a press event in New York. The personal computer was now respectable, legitimized by IBM's prestige and reputation. Demand was overwhelming and IBM immediately quadrupled production. Those in doubt were reassured. The IBM PC was sold through existing retail outlets not by IBM's expensive sales force, except as an add-on office accessory to larger systems, and advertising preserved the ambiguity as to whether the machine was for the home or office.

The growing user-base attracted software developers. The scope and variety of packages mushroomed. The PC was 'open architecture', i.e. specification and internal detail was made freely available. This allowed third parties to create cards to use in the expansion slots for graphics, sound, additional memory, networking, and additional ports for interfacing. The IBM PC became the industry standard. With no proprietary parts, a thriving business in clones arose and by the mid-1980s half of new PCs were compatibles. Compaq and Dell are two of many companies that successfully exploited the PC-platform clone market.

The success of the PC catapulted none other than Microsoft into the big-time. IBM needed system software for the PC and approached Digital Research that had developed the popular CP/M operating system software. For reasons that remain obscure (but colourful nonetheless) the negotiations failed and Microsoft inherited the offer. It is rumoured that Gates and Allen, exceptionally, wore suits for the meeting with IBM when the two teams met at Microsoft's rented premises in Seattle in July 1980. A deal was struck and Microsoft supplied the operating software, MS-DOS (Microsoft Disk Operating System). Just about every PC and PC clone had MS-DOS bundled with it, and Microsoft received a royalty for each and every licensed copy. By the mid-1990s there were over fifty million PC-platform computers running Microsoft software. Fate was capricious. Microsoft seized the day.

Chapter 7
The future of history

Futures past were not foreseen. From the start of automatic computing, predictions of what was to be were dismally bad. Bafflingly so in the light of what came to pass. Can we do any better with the future of history?

The collapse of categories

The history of artefacts offers us an aerial view of a larger predicament. Earlier I described the River Diagram with four threads: calculation, automatic computing, information management, and communication. From the times of knotted cords and abacuses, form was an indicator of function. What a device did was related to what it looked like. Devices in one thread were distinct from devices in another. A slide rule does not look like a radio. An arithmometer does not look like a speaking tube. Objects within a given thread were also distinct. A telephone does not look like a fax machine. It was reliably the case that what was functionally distinct was also physically distinct. Until the arrival of semiconductors and integrated circuits.

As the number of transistors on a microchip increased rapidly from the 1970s, doubling every two years, devices with multiple functions began to appear, and longstanding categories began to break down. How do we categorize a smartphone? Camera, address book, alarm clock, compass, torch, map, radio, television, messager, media player, calendar, calculator, watch...? Oh yes, a telephone. You can make and receive calls with it. The device looks like none of the artefactual precedents for each of its separate functions. How do we classify a tablet or laptop? Semiconductor technologies acted as a fusion chamber with a wide scatter of new outcomes. The result was the collapse of familiar object categories defined by use.

Is the collapse of categories a crisis for history? No, it *is* history. Computers are artefacts of our own making. The smartphone is a computer platform, an increasingly prominent one. While the variety of functions may be dazzling, each is a program written by a person or persons, with intention, skills, aspirations, and knowledge. Their values, social and professional contexts inform them, consciously and unconsciously, what is potentially desirable to have. Everything new is in the context of the past. The telephone and satellite networks that link machines have histories. The smartphone can be seen as a 'universal' machine that can run any program appropriately expressed. The history of computers as universal machine goes back at least to the 1930s and is well studied. Should it irk us that history does not yet have reassuring perspectives to frame such transformative changes? History is playing catch-up. It wouldn't be history if it wasn't. It would be news.

Decentring

The history of computing has, for most of its life, been machine-centred, with the computer the focus of its own story, largely estranged from broader cultural studies. This is changing. Media studies has newly engaged with computer history. Arts and humanities faculties contemplate how to deploy digital technologies and new conceptual models in ways relevant to their own fields of study. History is seen as a way for broader disciplines to better understand the implications and potential of digital technologies to their own domains of interest. Perspectives have been shifting from machine to context, and the machine is no longer everywhere the sole focus.

Decentering the computer is literal as well as interpretative. We are used to thinking of computers as finite localized objects. Monstrous mainframes, fridge-sized minicomputers, desktop personal computers, portable laptops, smartphones—single entities bounded in space. But the machine is diffusing into our environments. In a networked world a computer is one of countless interconnected machines. Micro-controllers are embedded in home appliances—microwave ovens, dishwashers, bread-makers. The 'Internet of Things' (IoT) refers to a network of devices with internal sensors, software, and communications capability to exchange data and orchestrate their operation—security, central heating, lighting—controllable, if needs be, from a smartphone or laptop. The 'Smart Home'. Micro-computers manage our cars' performance. As the computer becomes less visible it ceases to be the main protagonist in its own narrative. Increasingly, there will be histories that are less about the machines and more about the context and purpose of their use.

Software

I have not said much about software. This reflects, in an exaggerated way, the fractured attention it commands in histories of computing.

Software is a collective name for the programs that direct the computer what to do. The term came into use in the 1950s and makes a distinction between programs, on the one hand, and the physical hardware of computer systems, on the other. The two work in intimate union. They are inseparable parts of a functioning system.

Software is not integrated into histories of computing, and the apparent independence of the two histories is puzzling and unsatisfying. Software did not exist seventy years ago. It is an artefact like no other. There is art and craft, as well as exacting compliance with uncompromising rules. It is both brittle and flexible.

To understand a physical artefact we might ask how it is made. How programs are 'made' does not comply with models of engineering design or industrial manufacture. Putting more programmers on a project could actually make things worse. IBM found this to its great cost in the creation of an operating system for the System/360 in the 1960s. Massive financial and delivery overruns and a desperation to meet the near-ruinous challenges of producing so intricate and complex a product were chastening lessons for the industry.

'Software engineering' was an initiative in the late 1960s with an aspirational agenda to systematize software production using established engineering disciplines as the model. The initiative was a response to the 'software crisis'—the difficulty of managing the creation of large complex software systems that were reliable, maintainable, and delivered on time and to budget. Disciplining the process would, it was hoped, corral a new breed of creative (unruly) programmers by professionalizing software practices, effectively deskilling programming by proceduralizing it. The initiative had indifferent results.

We do not have a model for software as a human activity or as product, so its history is difficult to frame. The historian Michael Mahoney wrote in 2005 that 'historians of computing have only begun to tackle the history of software. We've stuck pretty close to the machine.' Early histories of computing tended to be technical descriptions of hardware systems written by pioneers and practitioners. This broadened out in due course into business, social, and economic histories, and this expansion continues. The history of software followed a similar trajectory but a decade or two in arrears. It started with histories of programming languages and coding practices, and, in the 1980s and 1990s, began tentatively to broaden, but not to the extent of hardware histories that were already liberalizing.

The history of software and of the software industry is growing but remains fragmented and incomplete. Programming languages are best covered. There is then a taper when it comes to histories of operating systems and databases and, at the thin end, applications software. That histories of applications software might be stirring, signals an emergence from technical introspection. Encouragement would not go amiss. Software is, after all, what determines what any given machine does. When it comes to the computer's agency in the world, the history of computing is increasingly the history of software, and software has yet to take its proper place.

A new master narrative

Framing the computer as a scientific tool for mathematical calculation reflects the intention and use to which computers were first put. Framing the computer as an 'information machine' reflects its use in the office automation era for business, administration, and data processing. Science and business. Each narrative marshals a set of canonical machines as devices of reference, or 'identifiers', illustrative and defining.

The machine of reference for online real-time computing is SAGE, the US air defence network built during the Cold War in response to the threat of Soviet airborne attack. A technological answer to a political problem. The Cold War was not the only context for online real-time systems. Totalisators, vast electromechanical systems for managing betting at dog and horse tracks, are capable of real-time computation in live response to the continuously changing placement of bets. The system could process bets from hundreds of simultaneously operated ticket machines, could automatically regulate peak traffic loads and spikes in demand, and was failsafe. Automatic Totalisators pre-date SAGE by several decades. It is SAGE that is enshrined in the canon. Totalisators are mentioned fleetingly, if at all.

When it comes to canonical honours it would seem that history privileges electronic solutions, understandably so, given that it is the electronic computer that is the focus of modern historical attention. But I do not believe that this is sufficient on its own to account for the Totalisator being blanked. Totalisators were not created as machines for scientific calculation or administration. They are machines for leisure, entertainment, sport, or gambling, and these fall outside the defining scope of the two established narratives.

The 'computer as information machine' is an existential description that may well be durable beyond its office data-processing origins. But this depiction does not capture the ingression of computers into arenas of human activity that are new—gaming, entertainment, sport, media, graphics, communication platforms. It is not that the Totalisator is insignificant, but rather that events have outgrown our narratives. Simulated sports, video gaming, and the growth of online gambling are outside the timeframe of our two main narratives. Whatever it is that the Totalisator was a precursor to does not yet feature in our histories. So it has no place of its own, even as a backstory for what came to be.

What we can look forward to is a response to the need for a new, or at least an updated, story that embraces the variety of new activities, contexts, and communities that are the machines' new hosts.

Further reading

Chapter 1: History and computing

Stan Augarten, *Bit by Bit: An Illustrated History of Computers* (George Allen & Unwin, 1984).
David Edgerton, *The Shock of the Old: Technology and Global History since 1900* (Profile Books, 2006).
James Gleick, *The Information: The History, the Theory, the Flood* (Fourth Estate, 2011).
Thomas Haigh, ed., *Histories of Computing: Michael Sean Mahoney* (Harvard University Press, 2011).
David Link, *Archaeology of Algorithmic Artefacts* (University of Minnesota Press, 2016).

Chapter 2: Calculation

James W. Cortada, *Before the Computer* (Princeton University Press, 2000).
Matthew L. Jones, *Reckoning with Matter: Calculating Machines, Innovation and Thinking About Thinking from Pascal to Babbage* (University of Chicago Press, 2016).
Florin-Stefan Morar, 'Reinventing Machines: The Transmission History of the Leibniz Calculator'. *British Journal for the History of Science (BJHS)* 48:1.1 (2015): 125–46.
Michael R. Williams, *A History of Computing Technology* (Prentice-Hall, 1985).

Chapter 3: Automatic computation

William Aspray, ed., *Computing before Computers* (Iowa State University Press, 1990).
Bernard I. Cohen, 'Babbage and Aiken'. *Annals of the History of Computing* 10.3 (1988): 171–91.
Ronald R. Kline, 'Inventing an Analog Past and a Digital Future in Computing'. *Exploring the Early Digital*. Ed. Thomas Haigh (Springer, 2019): 19–39.
Sydney Padua, *The Thrilling Adventures of Lovelace and Babbage: The (Mostly) True Story of the First Computer* (Penguin, 2015).
Brian Randell, 'From Analytical Engine to Electronic Digital Computer: The Contributions of Ludgate, Torres, and Bush'. *Annals of the History of Computing* 4.4 October (1982): 327–41.
Doron Swade, *The Cogwheel Brain: Charles Babbage and the Quest to Build the First Computer* (Little, Brown, 2000).
Doron Swade, 'Turing, Lovelace, and Babbage'. *The Turing Guide*. Eds. Jack Copeland et al. (Oxford University Press, 2017): 249–62, 506–8.

Chapter 4: Electronic computing

Martin Campbell-Kelly, William Aspray, Nathan Ensmenger, and Jeffrey R. Yost, eds. *Computer: A History of the Information Machine*. 3rd ed. (Westview Press, 2014).
James W. Cortada, *Before the Computer: IBM, NCR, Burroughs, & the Industry They Created, 1865–1956* (Princeton University Press, 2000).
Thomas Haigh and Mark Priestley, 'Colossus and Programmability'. *IEEE Annals of the History of Computing* 40.4 (2018): 5–27.
Thomas Haigh, Mark Priestley, and Crispin Hope, 'Reconsidering the Stored-Program Concept'. *IEEE Annals of the History of Computing* 36.1 (2014): 4–17.
Thomas Haigh, Mark Priestley, and Crispin Hope, *ENIAC in Action: Making and Remaking the Modern Computer* (MIT Press, 2016).
Maurice Wilkes, *Memoirs of a Computer Pioneer* (MIT Press, 1985).
Michael R. Williams, 'The Origins, Uses, and Fate of the EDVAC'. *IEEE Annals of the History of Computing* 15.1 (1993): 22–38.

Chapter 5: The computer boom

Morton M. Astrahan and John F. Jacobs, 'History of the Design of the SAGE Computer—The AN/FSQ-7'. *Annals of the History of Computing* 5.4 (1983): 340–9.
Martin Campbell-Kelly, *From Airline Reservations to Sonic the Hedgehog: A History of the Software Industry* (MIT Press, 2004).
Paul E. Ceruzzi and Thomas Haigh, *A New History of Modern Computing* (MIT Press, 2021).
James W. Cortada, *IBM: The Rise and Fall and Reinvention of a Global Icon* (MIT Press, 2019).

Chapter 6: Revolution

David C. Brock, and David A. Laws, 'The Early History of Microcircuitry: An Overview'. *IEEE Annals of the History of Computing* 34.1 (2012): 7–19.
Tracy Kidder, *The Soul of a New Machine* (Little, Brown 1981).
Douglas K. Smith and Robert K. Alexander, *Fumbling the Future: How Xerox Invented, Then Ignored, the First Personal Computer* (toExcel, 1999).

Chapter 7: The future of history

Martin Campbell-Kelly, 'The History of the History of Software'. *IEEE Annals of the History of Computing* 29.4 (2007): 40–51.
Doron Swade, 'Forgotten Machines: The Need for a New Master Narrative'. *Exploring the Early Digital*. Ed. Thomas Haigh (Springer, 2019): 41–68.

Index

Note: Tables, figures, and boxes are indicated by an italic following the page number.

For the benefit of digital users, indexed terms that span two pages (e.g., 52–53) may, on occasion, appear on only one of those pages.

C

D

E

F

G

H

I

J

K

L

M

N

O

P

Q

R

S

T

U

V

W

Z

INFORMATION

A Very Short Introduction

Luciano Floridi

Luciano Floridi, a philosopher of information, cuts across many subjects, from a brief look at the mathematical roots of information - its definition and measurement in 'bits'- to its role in genetics (we are information), and its social meaning and value. He ends by considering the ethics of information, including issues of ownership, privacy, and accessibility; copyright and open source. For those unfamiliar with its precise meaning and wide applicability as a philosophical concept, 'information' may seem a bland or mundane topic. Those who have studied some science or philosophy or sociology will already be aware of its centrality and richness. But for all readers, whether from the humanities or sciences, Floridi gives a fascinating and inspirational introduction to this most fundamental of ideas.

'Splendidly pellucid.'

Steven Poole, The Guardian

www.oup.com/vsi